MANUFACTURING TECHNOLOGY RESEARCH

ADDITIVE MANUFACTURING

OPPORTUNITIES, CHALLENGES, IMPLICATIONS

MANUFACTURING TECHNOLOGY RESEARCH

Additional books in this series can be found on Nova's website under the Series tab.

Additional e-books in this series can be found on Nova's website under the e-book tab.

MANUFACTURING TECHNOLOGY RESEARCH

ADDITIVE MANUFACTURING

OPPORTUNITIES, CHALLENGES, IMPLICATIONS

WILLIAM H. PHILLIPS
EDITOR

nova publishers
New York

Library of Congress Cataloging-in-Publication Data

ISBN: 978-1-63484-232-7

Published by Nova Science Publishers, Inc. † New York

CONTENTS

PREFACE

Additive manufacturing has the potential to fundamentally change the production and distribution of goods. Unlike conventional or subtractive manufacturing processes, such as drilling, which create a part by cutting away material, additive manufacturing builds a part using a layer-by-layer process. Additive manufacturing has been used as a design and prototyping tool, but the focus of additive manufacturing is now shifting to the direct production of functional parts—parts that accomplish one or more functions, such as medical implants or aircraft engine parts—that are ready for distribution and sale. This book discusses the use of additive manufacturing to directly produce functional parts, including its opportunities, key challenges, and key considerations for any policy actions that could affect its future use. This book explores the development and application of additive manufacturing as well as initiatives in the United States and other countries to advance it. It also examines the technology's effect on firm and industry production activities, as well as the potential implications for U.S. manufacturing competitiveness focused in three industries.

In: Additive Manufacturing
Editor: William H. Phillips
ISBN: 978-1-63484-232-7

Chapter 1

3D PRINTING: OPPORTUNITIES, CHALLENGES, AND POLICY IMPLICATIONS OF ADDITIVE MANUFACTURING*

U.S. Government Accountability Office

HIGHLIGHTS OF A FORUM CONVENED BY THE COMPTROLLER GENERAL OF THE UNITED STATES JUNE 2015

WHY GAO CONVENED THIS FORUM

Additive manufacturing has the potential to fundamentally change the production and distribution of goods. Unlike conventional or subtractive manufacturing processes, such as drilling, which create a part by cutting away material, additive manufacturing builds a part using a layer-by-layer process. Additive manufacturing has been used as a design and prototyping tool, but the focus of additive manufacturing is now shifting to the direct production of functional parts— parts that accomplish one or more functions, such as

* This is an edited, reformatted and augmented version of a report, GAO-15-505SP, issued by the United States Government Accountability Office, June 2015.

medical implants or aircraft engine parts—that are ready for distribution and sale.

On October 15-16, 2014, GAO, with the assistance of the National Academies, convened a forum to discuss the use of additive manufacturing to directly produce functional parts, including its (1) opportunities, (2) key challenges, and (3) key considerations for any policy actions that could affect its future use.

Forum participants included officials from government, business, academia, and nongovernmental organizations that were selected to represent a range of viewpoints and backgrounds. This report summarizes the ideas and themes that emerged during the forum and the collective discussions of the forum participants. This report does not necessarily represent the views of the organizations whose representatives participated in the forum, including GAO. Participants were given the opportunity to comment on a draft of this report and their technical comments were incorporated, as appropriate.

WHAT PARTICIPANTS SAID

Forum participants identified many opportunities for using additive manufacturing— also known as three-dimensional (3D) printing—to produce functional parts and discussed benefits that have been realized in the medical, aerospace, and defense sectors. For example, they said that the medical industry is using additive manufacturing to produce customized prosthetics and implants, including cranial implants. Because it is made specifically for a patient, the part results in a better fit, which leads to a better medical outcome. In the aerospace industry, participants said additive manufacturing was used to design and produce a complex jet engine fuel nozzle as a single part, which will reduce assembly time and costs for the engine. Participants identified some future applications of additive manufacturing including enhancing supply chain management. Overall, participants concluded that additive manufacturing will not replace conventional manufacturing, but rather it will be an additional tool for manufacturers to use when it is appropriate from a cost-benefit perspective.

Forum participants identified three broad groups of challenges in using additive manufacturing to produce functional parts: (1) ensuring product quality, (2) limited design tools and workforce skills, and (3) supporting increased production of functional parts. First, they identified challenges related to building quality parts, such as the need to improve the quality control of the additive manufacturing process. Second, they said that existing design and analytical tools combined with an insufficiently skilled workforce could limit the use of additive manufacturing and its ability to reach its potential for greater innovation. Finally, participants identified challenges that affect the increased production of functional parts, such as the need for an improved industrial infrastructure, including more robust supply chains of machines and materials.

Forum participants identified key considerations for potential federal policy actions that could affect the future use of additive manufacturing, including industry challenges, areas affected by additive manufacturing growth, and trade-offs. Although there was no consensus on specific policy actions needed and many participants suggested caution on potential government action, participants discussed several areas of potential government involvement, such as coordinating standards setting, considering risks for infringement of intellectual property rights with regard to additive manufacturing products, and encouraging a national dialogue about the government's role and its goals.

* * *

June 24, 2015

The Honorable Lamar Smith
Chairman
Committee on Science, Space, and Technology
House of Representatives

Dear Mr. Chairman:

Additive manufacturing—also known as three-dimensional (3D) printing— has the potential to fundamentally change the structure and management of supply chains and the production and distribution of goods. As the technology has matured, the use of additive manufacturing has become more widespread. Additive manufacturing has been used as a design and prototyping tool since its inception, but the focus of additive manufacturing is now shifting to the production of functional parts and products, such as jet engine parts and hearing aids.

On October 15-16, 2014, GAO, with the assistance of the National Academies, convened a select group of leaders and experts for a forum to discuss key issues related to using additive manufacturing for production of functional parts. These participants included representatives of government, business, academia, and nongovernmental organizations and were selected to represent a range of viewpoints and backgrounds. The forum offered a series of facilitated discussions designed to address the following questions:

1) What are the opportunities to produce functional parts using additive manufacturing?
2) What are the key challenges, if any, to using additive manufacturing to produce functional parts?
3) What are the key considerations for any policy actions that could affect the future use of additive manufacturing to produce functional parts?

This report summarizes the ideas and themes that emerged during the forum and the collective discussions of the forum participants. The report does not necessarily represent the views of the organizations whose representatives participated in the forum, including GAO. Participants were given the opportunity to comment on a draft of this summary report.

Appendix I provides more details on the scope and methodology to plan and conduct the forum and to prepare this report. Appendix II provides a list of the forum participants, and appendix III provides the forum agenda. Appendix IV provides a summary of the participant-identified challenges to using additive manufacturing to produce functional parts.

We conducted our work from March 2014 to June 2015 in accordance with all sections of GAO's Quality Assurance Framework that are relevant to our objectives. The framework requires that we plan and perform the engagement to obtain sufficient and appropriate evidence to meet our stated objectives and to discuss any limitations in our work. We believe that the information and data obtained, and the analysis conducted, provide a reasonable basis for any findings and conclusions in this product.

I would like to thank the forum participants for taking the time to share their knowledge, insights, and perspectives on this important and strategic national topic.

Sincerely yours,
Gene L. Dodaro
Comptroller General of the United States

FOREWORD

Additive manufacturing has existed for decades, but is now receiving increased attention because of its potential to transform the manufacturing industry and how goods are produced, distributed, and sold to consumers. For instance, additive manufacturing could make existing product supply chains more efficient by allowing for more on demand production, which could reduce the need to maintain large product inventories and spare parts, and allow for the localized production of goods closer to consumers. The do-it-yourself or maker movement has already benefited from the increasing availability of low-cost 3D printers and uses additive manufacturing technology to create a wide variety of items, including jewelry, toys, sculptures, and other artistic products. Additive manufacturing holds the potential for disrupting existing and creating new markets, but the technology is in its relative infancy and it may be years or decades before it reaches levels of confidence comparable to what the industry has with the more familiar conventional manufacturing processes and materials.

We focused this forum on the opportunities, challenges, and policy implications of the use of additive manufacturing for functional part production. As opposed to artistic products that may only be aesthetically pleasing, functional parts have structural and performance requirements to accomplish the function it needs to perform. Such parts, which could include items as simple as knobs or as complex as automobile or jet engine parts, need to be reliably and consistently produced even as the context for safety and performance can vary widely. As a result, there could be significant problems with poorly manufactured functional parts, particularly those used for safety-critical applications, such as parts for aircraft or medical products. Nevertheless, the potential benefits of using additive manufacturing for part design and production include the ability to fabricate more complex shapes, reduced manufacturing cycle times, costs, and supply chain risks, and more efficient use of manufacturing materials, which could potentially reduce waste in the manufacturing process and its resulting environmental effects.

However, the role of government policy making for additive manufacturing needs to be better understood. Determining that role will require a better understanding of the opportunities afforded by additive manufacturing for functional part production and the challenges currently faced. Further, as the use of additive manufacturing increases, particularly for safety-critical parts and products, it will be important to understand any potential concerns it raises.

I am grateful to all the forum participants for their time and insights during the meeting and throughout the development of this report.

Sincerely yours,
Timothy M. Persons, Ph.D.
Chief Scientist
Director, Center for Science, Technology, and Engineering

INTRODUCTION

Additive manufacturing is a layer-by-layer process of producing 3D objects directly from a digital model.[1] As seen in Figure 1, unlike conventional or subtractive manufacturing processes, such as drilling or milli, which create a part or product by cutting away material from a larger piece, additive manufacturing builds a finished piece in successive layers, generally without the use of molds, casts, or patterns, which can potentially lead to less waste material in the manufacturing process.

While the concepts have existed for decades, commercialization of additive manufacturing began in the mid-1980s and its first uses were primarily for presentation purposes. For more than 20 years, the technology has been evolving as a design and prototyping tool. Additive manufacturing offers the ability to rapidly create prototypes that can help validate the fit, form, and functionality of proposed products, which has provided both great time and cost savings in the product development cycle. As the technology has matured, the use of additive manufacturing has become more widespread and has expanded into more applications. For instance, one of the significant applications for additive manufacturing has been the production of tools and casts for conventional manufacturing. Lower manufacturing tool costs have allowed manufacturers to produce in lower volumes that previously may not have been cost-beneficial.

The use of additive manufacturing for prototyping and manufacturing tooling has helped to improve the efficiencies of conventional manufacturing processes, and the use of additive manufacturing is now shifting to the direct production of goods that are ready for distribution and sale. The emergence of desktop equipment for additive manufacturing has enabled the production of jewelry, art replicas, toys, models, and other artistic products.

Subtractive manufacturing

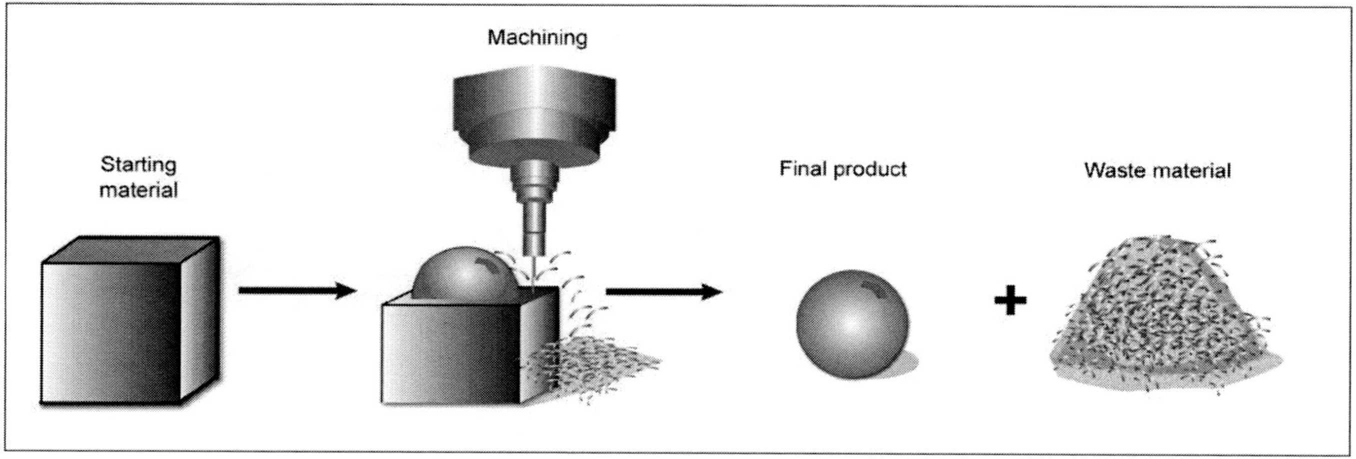

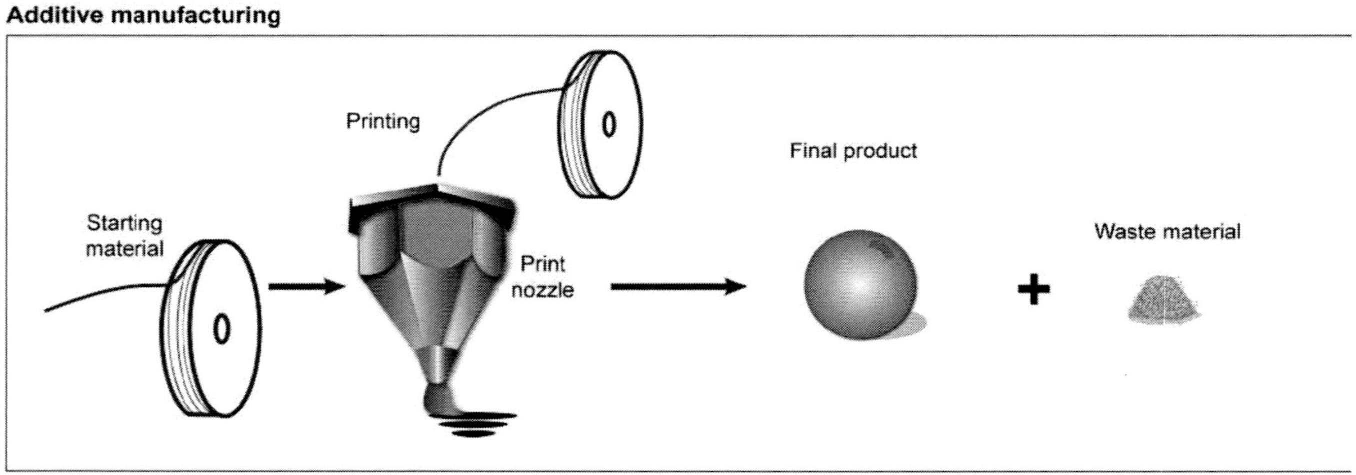

Source: GAO (analysis); Art Explosion (images). | GAO-15-505SP.

Figure 1. Conceptual Comparison of Subtractive and Additive Manufacturing.

However, it is the potential to use additive manufacturing to produce functional parts and products, particularly in critical applications such as medicine and aerospace, that has generated a lot of attention. For instance, additive manufacturing has been described as having the potential to transform the manufacturing industry and lead to a "new industrial revolution," a term used by The Economist in 2011.[2] In 2012, the National Defense University said that additive manufacturing was a prospective game-changer that could be part of a U.S. manufacturing revolution.[3] In July 2014, Gartner, a leading information technology research firm, identified consumer and industrial applications of 3D printing, among others, as having transformational business benefits, but estimated that mainstream adoption would not occur for 5 to 10 years.[4]

The manufacturing industry's adoption of additive manufacturing to directly produce goods is in its infancy but growing rapidly. In 2011, the National Institute of Standards and Technology (NIST) estimated that the United States produced an estimated $246.1 million from additively manufactured goods, which represented about one-hundredth of 1 percent of all manufactured goods produced in the United States.[5] In the 2-year period from 2011 to 2013, Wohlers Associates, an independent consulting firm that specializes in additive manufacturing, estimated that the global revenue from additively manufactured goods grew more than 65 percent from $642.6 million in 2011 to $1.065 billion in 2013.

Several Types of Additive Manufacturing Processes and Technologies Exist

Additive manufacturing is not a single type of technology or process. All additive manufacturing systems employ a general layer-by-layer approach, but they use a wide variety of technologies, materials, and processes. For example, materials used in additive manufacturing include plastic, metal powders, ceramics, and foundry sand. In 2012, ASTM International defined seven categories of additive manufacturing processes to group the different types of technologies (see table 1).[6]

Table 1. Types of Additive Manufacturing Processes

Process name	Description
Binder jetting	A liquid bonding agent is selectively deposited to join the powder materials.
Directed energy deposition	Focused thermal energy, such as a laser, is used to fuse materials by melting as the materials are being deposited to form an object.
Material extrusion	Materials are heated and selectively dispensed through a nozzle.
Material jetting	Materials, such as photopolymers or wax, are selectively dispensed through a nozzle.
Powder bed fusion	Thermal energy selectively fuses regions of a powder bed.
Sheet lamination	Sheets of materials are cut and stacked to form an object.
Vat photopolymerization	The use of certain types of light, such as ultraviolet light, to selectively solidify liquid photopolymers.

Source: GAO analysis of ASTM International data. | GAO-15-505SP.

Note: Photopolymers are materials that transform from liquid to solid in the presence of certain types of light.

The categories of additive manufacturing systems have similarities, but there are differences in their capabilities that would affect their usage for different applications or markets.

For instance, powder bed fusion and directed energy deposition both use similar energy sources, but one melts material on a powder bed, while the other melts material while depositing.

Even within each category, additive manufacturing systems have different levels of capability or use different technologies. For instance, different additive manufacturing systems vary in the materials they can utilize, the speed at which they can build parts, the dimensional accuracy and quality of the surface finish of the produced parts, the material properties of the produced parts, costs of materials and machines, accessibility and safety related to complexity of operation, and other capabilities such as multiple colors.

For example, Figure 2 illustrates two additive manufacturing processes, powder bed fusion and material extrusion. Powder bed fusion, here illustrated by laser sintering, can be used to produce parts from powdered materials using a laser to selectively fuse or melt the particles at the surface, layer by layer, in an enclosed chamber.

The patent for laser sintering was originally filed in 1986 and forms the basis for current powder bed fusion methods. Powder bed fusion can use a variety of materials, including plastic, metal, ceramic, and glass powders, to create parts.

Essentially, for each layer of the part being produced, a new layer of powder is presented to the laser. The areas that are targeted by the laser are based on each layer of the 3D digital model of the part. Machines for these processes come in a variety of sizes, with smaller units having a footprint of 2 to 3 feet wide and deep and a height of 7 feet, and larger units about the size of a small compact car. Small powder bed fusion machines can cost tens of thousands of dollars; those designed for production can range from hundreds of thousands of dollars, up to a million dollars or more. An example of a part produced by powder bed fusion using metal powder is an aircraft seat belt buckle.

An example of material extrusion is fused deposition modeling. Here, materials are heated and melted through an extruder, which acts like an automated version of a hot-glue gun used for arts and crafts. The extruder moves horizontally and vertically to deposit the material through a nozzle layer by layer.

Fused deposition modeling was originally patented in 1989 and primarily uses plastic filaments as a material, but the use of metal wire as a material has been researched. Material extrusion machines range from desktop-sized design units (approximately 2 feet wide and deep) to production or large prototyping machines, which are approximately the size of a compact car.

Desktop machines can be obtained for anywhere from a few hundred dollars to tens of thousands of dollars, and prototype or production machines can cost tens or hundreds of thousands of dollars. An example of a part produced by material extrusion is a taillight for an automobile. Today, most of the additive manufacturing equipment that is widely available for in-office and home environments is based on material extrusion, which has helped to contribute to the do-it-yourself or maker movement (see Figure 3).

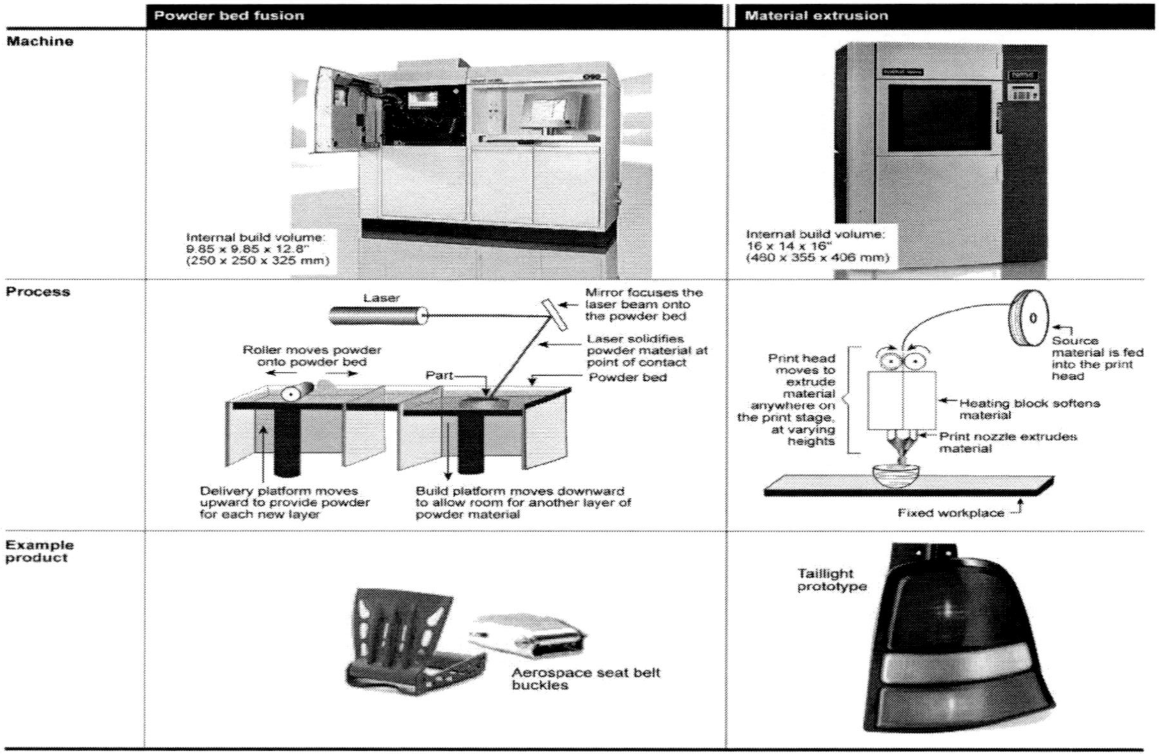

Source: EOS GmbH (left top and bottom images); GAO (middle left and right images) and Stratasys (right top and bottom images).

Figure 2. Examples of Additive Manufacturing Systems. | GAO-15-5-5SP.

Source: 3D Systems. | GAO-15-505SP.

Figure 3. Desktop 3D Printer.

Applying Additive Manufacturing to the Part Production Process

For product manufacturers, the general process for using additive manufacturing may be similar to conventional manufacturing. The product manufacturer still needs to purchase equipment for its factories from equipment manufacturers and maintain the equipment in good operating condition. Raw materials to manufacture products are still needed. Product manufacturers will employ designers that use computer-aided design (CAD) and other related tools to design products that will be sold to their customers.

However, to use additive manufacturing, product manufacturers may need to design new or modify existing business processes to integrate additive manufacturing equipment into their product development process, including managing materials supplies that are compatible with their equipment. Refining these processes so that the manufacture of products using additive manufacturing systems can be cost-beneficial will be critical. As with other manufacturing technologies, the business success of additive manufacturing will be based on the ability of manufacturers to produce high-quality parts and products in a timely manner, at costs that customers are willing to pay, and that address customers' needs.

However, it is the potential ability to manufacture products outside of the factory that has contributed to the attention on additive manufacturing. As the cost of operating additive manufacturing systems decreases, and it becomes more feasible to operate and purchase such systems in multiple locations, it is possible that assembly lines and supply chains could be reduced for many parts and products. For instance, production could be moved closer to the consumer and possibly even be performed by consumers themselves. As a result, additive manufacturing could alter current structures related to production, packaging, distribution, and transportation. Further, because additive manufacturing enables the design and manufacture of custom products, it could drive the transition from mass production to mass customization, in which each produced item is customized for the consumer at little or no additional production cost.

The Federal Government's Involvement with Additive Manufacturing

Support from federal agencies, such as the National Science Foundation (NSF) and the Department of Defense (DOD), was instrumental in the early research and development into additive manufacturing. According to the Science and Technology Policy Institute, since 1986 when it first began funding additive manufacturing, NSF has expended more than $200 million on additive manufacturing research and related activities.[7] Now, several federal agencies are involved with the research and development of additive manufacturing, including NSF, the National Aeronautics and Space Administration (NASA), NIST, DOD, and the Department of Energy. Within DOD, several research organizations are involved, including the research laboratories of the Army, Navy, and Air Force and the Defense Advanced

Research Projects Agency (DARPA). These federal agencies support research at federal laboratories, academic institutions, and small and large companies, sponsor technical conferences, and participate in standards development. To help guide research and development efforts, federal research and development agencies have supported the development of several technology roadmaps. Further, in August 2012, the National Additive Manufacturing Innovation Institute, also known as America Makes, was founded as a public-private partnership to accelerate the research, development, and demonstration of additive manufacturing and transition technology to the manufacturing industry in the United States. Its federal partners include the Departments of Commerce, Defense, Education, and Energy, NASA, and NSF. America Makes is part of a broader National Network for Manufacturing Innovation that is designed to stimulate advanced manufacturing technologies and accelerate their commercialization in the United States. The interagency Advanced Manufacturing National Program Office manages the network and includes participation from all federal agencies involved in U.S. manufacturing. It is designed to enable more effective collaboration in identifying and addressing manufacturing challenges and opportunities that span technology areas and cut across agency missions.

Federal agencies are funding a wide variety of research and development activities, including the use of new materials, processes, and applications for additive manufacturing. For instance, DOD is researching the use of additive manufacturing technology to print living cells that can form biological tissue, such as skin for the treatment of burn victims and organs for transplants. It is also conducting research to print electronic components for circuitry or antennas. The Department of Energy and DARPA have established manufacturing demonstration facilities at the Oak Ridge National Laboratory and Pennsylvania State University, respectively, to facilitate the advancement and deployment of additive manufacturing technologies. NIST is funding research into measurement science to provide greater assurance into the quality of parts produced using additive manufacturing. It is also leading efforts on additive manufacturing standards through ASTM International's committee on additive manufacturing, which was formed in 2009. America Makes is funding research projects that are exploring the use of new materials, such as bioresorbable materials for medical devices and new metal alloys.[8]

Opportunities Exist to Use Additive Manufacturing for the Production of Functional Parts and for Other Applications

Forum participants identified many opportunities for using additive manufacturing to produce functional parts. They noted that the use of additive manufacturing has produced benefits, such as reduced time to design and produce functional parts, the ability to produce complex parts that cannot be made with conventional manufacturing processes, the use of alternative materials with better performance characteristics, and the ability to create highly customized, low-volume parts. Participants also identified some future applications of additive manufacturing, including greater use for building tooling for conventional manufacturing lines, enhancing education, and enhancing supply chain management. The maker community is also taking advantage of inexpensive 3D printing machines to produce modestly-sized polymer items. Overall, participants concluded that additive manufacturing will not replace conventional manufacturing processes, but it will be an additional tool for manufacturing parts to be used when it is appropriate from a cost-benefit perspective.

Current Applications of Additive Manufacturing Are Benefiting Industries

Forum participants identified several benefits that the aerospace, medical, and defense industries have realized from the use of additive manufacturing, including reducing the time to design and produce functional parts, whether the parts are manufactured using additive or conventional processes; producing complex parts with fewer subparts that cannot be made with conventional manufacturing processes; using alternative materials with better performance characteristics; and creating highly customized, low-volume parts—all of which can lead to cost savings.

Participants highlighted various uses of additive manufacturing by the medical industry, including customized prosthetics, implants, hearing aids, and dental applications. The aerospace industry is using additive manufacturing to produce several critical parts such as ducting for in-cabin air circulation, an aircraft fuel nozzle, and a satellite propellant tank. DOD is testing the use of additive manufacturing for prototype and temporary part manufacturing in remote locations, such as aboard ships and at forward operating bases.

According to forum participants, additive manufacturing can reduce the time to design and produce functional parts because it can produce prototypes rapidly without reconfiguring or retooling the manufacturing line, and it can provide the ability to implement new concepts, designs, and innovations quickly. Overall, the use of additive manufacturing can improve the speed of getting new products to market. For example, one participant described the development of a titanium satellite propellant tank made using electron beam wire fed additive manufacturing, as shown in Figure 4. Similar tanks produced using conventional manufacturing processes take about a year to design and manufacture. With additive manufacturing, it only takes about a month, reducing lead time and cost for customers. Another participant mentioned that additive manufacturing has been used for cranial implants because it can produce custom, patient-specific parts in 1 to 5 days. Using additive manufacturing also eliminates the upfront costs of tooling and molding, making it well-suited for such patient-specific parts.

Source: Lockheed Martin Corporation. | GAO-15-505SP.

Figure 4. Additively Manufacturing a Satellite Propellant Tank Using Electron Beam Technology.

Forum participants also described the savings from the use of additive manufacturing during the development cycle of parts produced with conventional manufacturing processes. For example, the Army's Rapid Equipping Force, which was created to provide immediate technology solutions to deployed forces, uses additive manufacturing to aid design through in-theater rapid prototyping in mobile Expeditionary Labs. These labs include additive manufacturing machines to enable Rapid Equipping Force teams to work with soldiers to quickly design and prototype solutions locally. Figure 5 shows an example of these labs. Additive manufacturing makes it easier to try a variety of designs and assess their field performance to optimize the design. For example, the Army has used this capability to design and prototype valve stem covers for a military vehicle, creating four iterations of designs to ensure fit and function, before using conventional manufacturing processes to create the final product.

Source: U.S. Department of Defense. | GAO-15-505SP.

Figure 5. Additive Manufacturing Systems in Expeditionary Labs.

Additive manufacturing allows manufacturers to produce some complex parts that cannot be made or are very expensive to make with conventional manufacturing processes, according to forum participants. This can enable manufacturers to create better designs that have fewer parts and use less material, which leads to reduced cost. For example, the aerospace industry is using additive manufacturing to create complex parts with fewer components than those made with conventional manufacturing. According to one participant, General Electric's (GE) use of additive manufacturing to produce jet engine fuel nozzles has allowed it to create a fuel nozzle as a single part rather than the 20 parts that would have been necessary using conventional manufacturing processes, which reduces the assembly time and costs because fewer brazes and welds are needed (see Figure 6).[9] Further, using additive manufacturing has enabled the fuel nozzles to be designed with more intricate cooling pathways and with five times more durability.

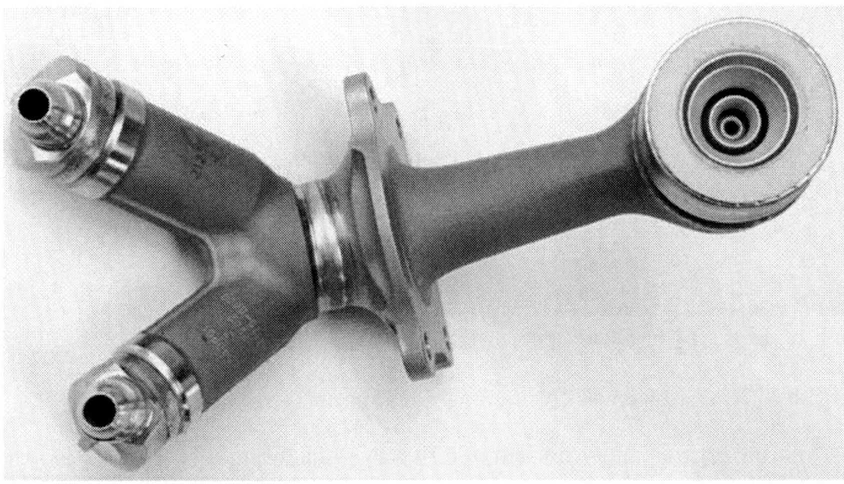

Source: General Electric. | GAO-15-505SP.

Figure 6. Jet Engine Fuel Nozzle.

Forum participants also described the use of additive manufacturing to create lighter polymer ducts for aircraft with shapes that cannot be machined or bonded together and provide a design advantage for air flow over conventionally manufactured ducts. In addition to a faster production time for the propellant tank for a satellite, participants said that using additive manufacturing has enabled a more complex tank design that can be larger than the current 47-inch diameter limitation that exists with forged die tooling. It also has fewer parts to assemble. These benefits have resulted in faster production and lower production and life-cycle costs for the satellite. Further, overcoming the forged part design limitation means that different sizes and geometries can be explored for future designs. For medical implants, complex internal mesh structures and porous structures enable bone to grow into the implant for better integration, as well as the ability to vary the density to more closely match the stiffness of the surrounding bone. Figure 7 shows an additively manufactured prototype tibial and femoral stem knee implant demonstrating the complex internal structure and porous surface achievable using additive manufacturing along with an x-ray of a knee implant in place to show how they would be used.

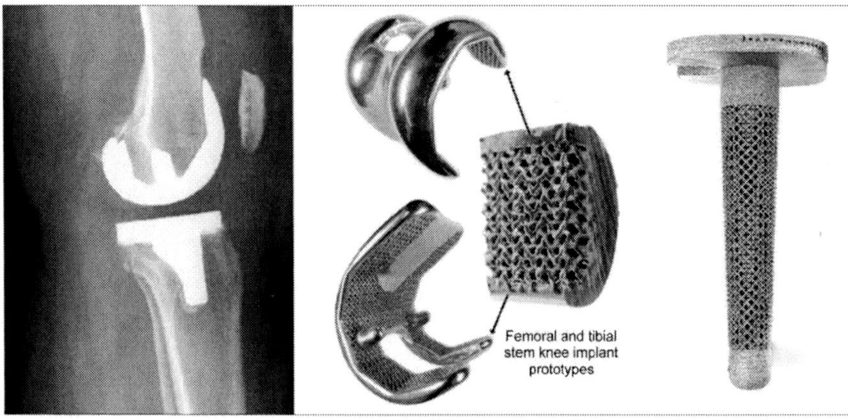

Source: Dr. L. E. Murr, University of Texas at El Paso. | GAO-15-505SP.

Figure 7. Complex Internal Structure of a Knee Implant.

Forum participants said that additive manufacturing also enables the use of alternative materials for parts that may result in better performance than equivalent parts manufactured using conventional materials, such as lower weight and greater strength. Several participants identified examples in the aerospace industry of lighter parts when using additive manufacturing instead of conventional manufacturing. For instance, one participant described the substitution of polymers for metal in aircraft ducts that results in lighter parts, which, over the lifetime of the aircraft, will reduce operating costs as a result of reduced fuel consumption. While using additive manufacturing to create the parts may sometimes cost more, overall cost savings can be achieved through simplified assembly due to fewer parts and reduced production and repair times. One participant provided an example of an airline using additive manufacturing to produce an arm to hold video monitors for its aircraft. While it cost more to produce these parts compared with using conventional processes, the parts incorporated an internal lattice structure, resulting in less material used and, therefore, less weight, resulting in an estimated savings in fuel use of more than $1.5 million over the life of the aircraft. Using additive manufacturing to create the aircraft engine fuel nozzles has enabled the use of an alloy material that is light, strong, and corrosion resistant and can operate at a 20 percent higher temperature than their conventionally manufactured counterparts. These additively manufactured nozzles weigh 25 percent less than their predecessor parts, reducing fuel costs. Another participant described an example of a cranial implant, which has been approved by the Food and Drug Administration, using a high-performance plastic material that is lighter than conventional materials with similar density and stiffness to bone and unlike conventional materials, such as titanium and stainless steel, does not interfere with diagnostic imaging equipment. Figure 8 shows an example of a plastic cranial implant.

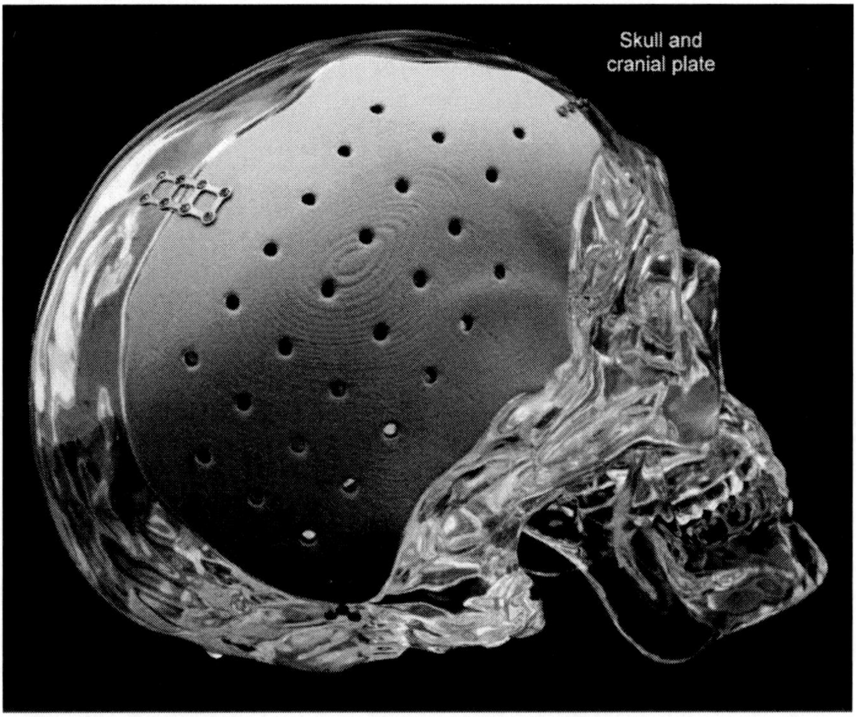

Source: Oxford Performance Materials, Inc. | GAO-15-505SP.

Figure 8. Custom Patient-specific Cranial Implant.

Forum participants said that one of the other key benefits of additive manufacturing is its use to create highly customized, low-volume parts quickly and at lower costs than using conventional manufacturing to produce comparable parts. For example, participants said that manufacturers in the medical industry are using it to produce customized prosthetics and implants faster and more cheaply. Using additive manufacturing enables each part to be made specifically for the individual patient's anatomy, which results in a better fit and an implant that is more structurally sound for a longer period of time, which, in turn, leads to better medical outcomes with fewer side effects. Prosthetics can also be customized for functionality to meet the patient's needs. Participants also stated that using additive manufacturing can result in easier and shorter surgeries. For instance, customized models and surgical guides enable surgeons to execute the treatment plan before the surgery, which reduces surgical time while also improving surgical outcomes.[10] One participant indicated that DOD has seen reductions of one-quarter to one-half of the time needed for surgery due to customization. Reducing the amount of time to perform surgeries can result in significantly reduced surgical costs, which, according to one participant, can be up to $300/minute.

Forum Participants Highlighted Some Future Uses for Additive Manufacturing

Forum participants indicated that, while the future use of additive manufacturing is uncertain, it is expected to expand in both existing and new applications, such as tooling for conventional manufacturing lines, enhancing education, and supply chain management. Additive manufacturing is already used for tooling, and some participants said that there is great potential for further growth to build tooling for conventional manufacturing lines, such as for stamping dies to make sheet metal parts for the automotive industry or to make custom molds for metal casting or plastic injection molding. However, one participant stated that, while additive manufacturing would work technically for producing tools, it would only be economical to produce high-performance tooling. Similar to functional part production, additive manufacturing enables the production of more complex tool designs with different geometric features incorporated into the tooling to change the characteristics of the end products, such as the surface finish or structure. Additionally, tools can be manufactured in less time and with reduced downtime for production, and the process may reduce the cost of tool production, and result in tools of higher quality.

Forum participants identified the use of additive manufacturing by teachers and instructors to enhance education as a key future use. The use of 3D printers to create teaching tools can improve lesson planning so that students can grasp concepts better and see real world applications of what they are learning and become more engaged. Additive manufacturing also can be used to create visual tools that enable students to visualize information in a way that is easier to comprehend and retain. New capabilities for 3D visualization of ideas could lead to a new generation of designers, engineers, scientists, and doctors.

According to forum participants, additive manufacturing will change supply chains, but it is not certain how or to what extent. It is unclear what parts will be printable in the future or how such manufacturing will affect the boundary between manufacturers and customers, as customers could become their own manufacturers for some types of parts. For instance, the maker community is growing exponentially and is already taking advantage of low-cost additive manufacturing machines to design and manufacture low-volume customized niche products and selling them over the Internet. It can also provide the capability to create mass-customized components, such as shoes, shells for cell phones, or other consumer items. Additive manufacturing enables parts to be built locally as needed, reducing the need for manufacturers to produce minimum lots and to ship and store material, which could make inventory forecasting more accurate and reduce financial constraints.

For example, participants stated that DOD is looking at ways to use additive manufacturing in supply chain management, including using it to repair equipment and produce parts in the field to reduce the need to store parts, to produce discontinued parts or temporary parts to use until a permanent part can be obtained, and to quickly build parts to meet mission requirements. This will enable personnel in the field to repair equipment, reduce equipment downtime, and execute their mission more quickly. Figure 9 shows some of the simple parts that could be additively manufactured in the field. One participant mentioned that one of the key DOD uses could be in supply depots and how inventory is managed in them and cited a prior GAO report on the topic. In December 2008, we found that the Navy had about $7.5 billion in inventory exceeding requirements,[11] and some participants indicated that the ability to additively manufacture parts on-demand could help reduce this inventory. Some participants stressed that additive manufacturing may reduce the need to stock spare parts, but it will not eliminate this need because many parts cannot be made through this process or cannot be made cost-effectively. The Army and Navy are exploring the use of additive manufacturing in the field and have experimented in making small parts, such as valve stem covers for vehicles and onboard logistics parts for ships that were difficult to locate such as drain and valve covers. The Marine Corps is just beginning to plan and evaluate the use of additive manufacturing.

According to one participant, during the Marine Corps's annual Expeditionary Logistics Wargame, a vendor demonstrated the use of additive manufacturing to repair small broken parts for an explosive ordinance robot using additively manufactured parts.

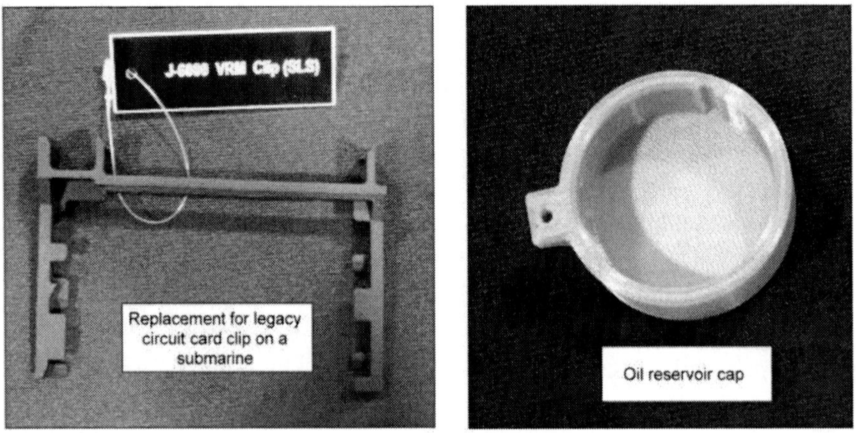

Source: U.S. Department of Defense. | GAO-15-505SP.

Figure 9. Simple Additively Manufactured Replacement Parts.

According to forum participants, while the use of additive manufacturing will grow in the future, it will not replace conventional manufacturing processes, but it will be an additional tool for manufacturing parts to be used, when appropriate, from a cost-benefit perspective. In many circumstances, conventional processes for manufacturing parts will always be faster, better, and more economical for extremely high-volume and well-proven parts. Additive manufacturing will likely be best for parts requiring properties and features not achievable through conventional manufacturing or for low-volume parts, such as custom items and tooling. One participant provided the example of producing titanium compressor blades for aircraft turbine engines. Millions of these blades are made every year using conventional processes, and additive manufacturing cannot compete with the efficiency and cost of these processes. Additive manufacturing can be integrated into the manufacturing process as a whole, such as using it for further processing of parts manufactured using conventional processes. Additive manufacturing can also be used to repair parts produced using conventional manufacturing processes. For example, one participant talked about using an additive manufacturing process called directed energy deposition—which enables the creation of parts by melting material as it is being deposited—to repair aircraft turbine blades rather than replacing them with new blades.

INDUSTRY FACES KEY CHALLENGES IN USING ADDITIVE MANUFACTURING TO PRODUCE FUNCTIONAL PARTS

Forum participants identified challenges in using additive manufacturing to produce functional parts. These challenges can be broadly grouped into three main areas: (1) ensuring product quality, (2) limited design tools and workforce skills, and (3) supporting increased production of functional parts, such as the need for a more robust industrial infrastructure. Appendix IV provides the full list of challenges identified by the participants during the forum meeting.

Additive Manufacturing Faces Challenges in Ensuring the Quality of Functional Parts

Forum participants identified the need to ensure the quality of functional parts, such as ensuring that manufacturers can repeatedly make the same part and meet exactness and consistent performance standards on the same machine or on different machines, as a key challenge. Additive manufacturing has historically been used to create prototypes that do not need to meet the same level of quality as final products and may not need to be exactly reproduced multiple times. However, as the technology is increasingly used to directly manufacture functional parts and competes with established manufacturing processes, the additive manufacturing industry needs to improve its ability to ensure product quality, according to the participants. This becomes particularly important for the medical, aerospace, and defense industries, where products must consistently meet high quality standards to be usable.

The quality of functional parts produced using additive manufacturing depends on a variety of factors. One participant highlighted several potential sources of variability that could affect product quality, including materials, machines, and manufacturing practices (see Figure 10). Current technology and practices do not ensure that these sources of variability can be consistently controlled to produce the desired product. For example, if manufacturers take a commercially available material (typically a powder) and use it in a commercially available machine, they would not be assured of consistently making the same product and having it consistently meet quality standards because variability in any of the elements of the manufacturing process could result in a final product that does not meet standards. For example, one participant said that these machines have different settings that can be used during production and that different manufacturers may set their machines up differently to create a part.

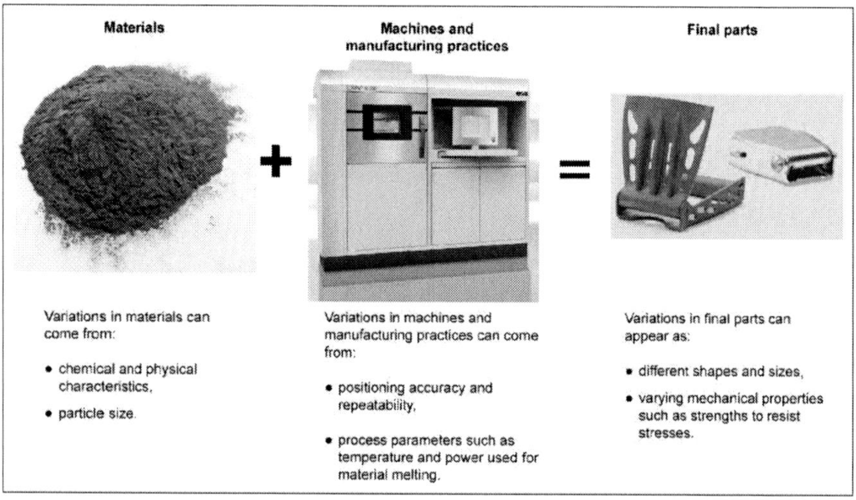

Figure 10. Examples of Variability in Additive Manufacturing Materials, Machines, Manufacturing Practices, and Final Parts.

Currently, additive manufacturing machines, particularly metal-based machines, can have difficulty maintaining the quality of final parts because there are sources of variability that affect the machines' ability to control the additive manufacturing process. For example, uneven temperature in certain types of additive manufacturing machines during production can lead to inconsistent dimensions and mechanical properties, such as strength, impact resistance, or hardness, in the resulting functional parts. These inconsistencies may lead to some parts not performing as intended. Another participant emphasized that the machines themselves are constantly evolving and improving, which creates another challenge in gaining control of the process. To address this challenge, participants said that manufacturers need better methods to control the additive manufacturing process to achieve high quality, including tools focused on measuring and controlling the additive manufacturing process. For example, a few participants mentioned the possibility of incorporating real-time monitoring systems into additive manufacturing machines to check the quality of each layer put down by the machine. One participant noted that some machines have limited monitoring systems now and that those systems could be expanded in the short term.

Forum participants said another challenge to ensuring the quality of functional parts produced is limited knowledge of additive manufacturing material properties. Specifically, manufacturers need to understand how a material will perform during processing and how the material's properties will affect the quality of the final part. Understanding material properties requires conducting research and developing material properties databases, which can be used to determine whether a part made from a specific material will meet quality requirements. For example, making products with high quality requirements, such as cranial implants that are customized to specific patients, requires a high degree of control of the material, according to one participant. Maintaining a high degree of control requires manufacturers to have a strong understanding of the material's underlying properties, as well as how variations in process influence the mechanical properties of the final part. As another participant explained, these material properties databases can provide the customer using the final product with a greater level of confidence and also help lower risks of using the product.

The additive manufacturing industry also needs to develop an improved suite of standards related to different elements of the additive manufacturing process, according to participants.[12] The industry has been developing a supporting body of knowledge for the technology over the past 3 decades, but that is not much time compared with conventional manufacturing processes, according to some participants. For example, the additive manufacturing industry has only been developing standards since about 2009, and several participants said that further standards are needed. One participant noted that the standards related to 3D building, including standards for geometric accuracy and surface characteristics, are not well-defined. Controlling machines to ensure reproducibility can take a great deal of time and expertise. With well-developed standards, manufacturers can have a more consistent approach to collect and report test data to ensure product quality. Moreover, standards development could help reduce variability in additive manufacturing processes, such as how machines are configured, maintained, and operated. Developing additional standards also could improve the confidence of product manufacturers, small- and medium-sized companies considering using additive manufacturing, and the customer using the product by providing a common understanding of how the product was manufactured and how it should perform. Such standards will be important to the development of the industry. However, one participant noted that the industry does not need to start from scratch because manufacturers will be able to make some use of the existing standards.

Participants stated that additive manufacturing needs to improve its ability to meet certification requirements. Certification by an independent body can be used to show that a product meets quality requirements and customers[13] or regulators[14] in certain industries may require that products meet specific standards. For example, additive manufacturing companies seeking to manufacture and sell safety critical airplane parts, such as engine parts, to aerospace companies must be able to have the manufacturing processes and parts certified. In order to facilitate this process, participants agreed that additive manufacturing needs consensus methods[15] and test data for qualification[16] of materials and production processes leading to certification of the final products.

Before a product can be certified, manufacturers must qualify the materials and processes used to make the part, which involves manufacturers conducting tests and collecting data under very controlled conditions. Other manufacturing processes have established methods and data based on a long history of use, but additive manufacturing does not have this support. For example, reliable databases of material properties are not generally available for additive manufacturing and are necessary for qualifying materials. By contrast, materials data to support other manufacturing processes, such as injection molding, are available to manufacturers. However, participants noted that qualifying new materials and processes is a time-consuming and expensive process, potentially taking several years. Improved consensus methods and test data could make the qualification process more efficient. In addition, one participant said that it can be challenging to qualify additive manufacturing processes because the machines are regularly being revised, and minor changes to a process can require requalification.

In order to certify the final product, manufacturers will need ways to test it to ensure that it meets established standards. Current testing methods are not always well-suited for additive manufacturing. For example, some commonly used testing methods destroy a given number of parts or test pieces to confirm that they meet established standards.[17] One participant noted that this type of testing requires destroying a lot of parts to ensure that those being manufactured for the customer have sufficient quality, which adds to the total costs of the final product. Another participant emphasized that developing better alternatives to these destructive tests have the potential to facilitate the certification process. In addition, improved nondestructive test methods could provide a more effective way to validate that a unique part, such as a cranial plate for a patient, was made correctly without having to destroy extras. Current inspection methods work very well for conventional products, but may not work as well with additive manufacturing, according to one participant. For example, specialized tools designed to measure the dimensions of a product may not fit as well into complex shapes or internal passageways and will not be as accurate for inspecting product quality.[18]

Additive Manufacturing Faces Challenges from Limited Design Tools and Workforce Skills

Forum participants said that existing design and analytical tools combined with an insufficiently skilled workforce could limit the use of additive manufacturing and its ability to reach its potential for greater innovation. As previously noted, building products with customized shapes is one of the potential benefits of using additive manufacturing compared with conventional methods (see Figure 11 for examples), but existing CAD software limits the ability of designers to explore new possibilities. For example, one participant noted that it is difficult for designers to fully use complex geometrical shapes such as ovals, mesh, or empty spaces using existing CAD software. CAD software designed to take advantage of additive manufacturing's capabilities would have clear advantages for designers, but one participant pointed out that there may not be enough incentive for CAD software developers to fully incorporate functionality to utilize all of the capabilities of additive manufacturing when it is a small part of the overall market.

Participants agreed that designers also will need tools capable of analyzing the products designed specifically for additive manufacturing to complement new CAD design tools. For example, if new CAD software allows designers to make use of complex geometrical shapes, tools are needed to analyze these new shapes. According to one participant, the product needs to be checked to show that it can meet functional requirements, such as whether the product has sufficient strength or resistance to impact. Another participant emphasized that analysis will be critical for any engineered products with specific quality requirements that are made with additive manufacturing. A challenge in developing analytical design tools involves developing the capability to measure and model the additive manufacturing process and then predicting how the process will affect the quality of the final product, according to participants. Specifically, designers require an improved understanding of how different elements of the additive manufacturing process, such as materials, machines, and manufacturing practices, affect product quality. In addition, analytical tools for additive manufacturing will need to integrate various components into a systematic approach to designing and engineering products for additive manufacturing including engineering analytical methods and test data on various material properties, one participant said. Developing this capability will require further research, according to participants.

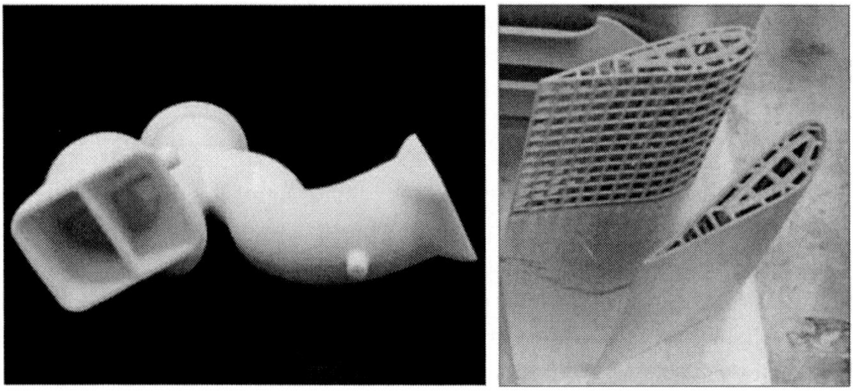

Source: 3D Systems (left) and Lyons, B., The Boeing Company (right). | GAO-15-505SP.

Figure 11. Examples of Additively Manufactured Products with More Complex Shapes (Aerospace Duct (left) and Trussed Air Foils (right)).

Participants also emphasized the value of developing a skilled workforce capable of thinking differently about design to take full advantage of additive manufacturing. Some participants pointed out the importance of having designers who can utilize the capabilities of additive manufacturing, such as building a product as one piece instead of building multiple pieces and assembling them into a product. One participant cited the GE jet engine fuel nozzle that the company decided to build using additive manufacturing. The initial design approach involved developing multiple CAD designs for separate components of the nozzle and then assembling the components, which is a traditional manufacturing approach. However, one engineer brought up the idea of manufacturing the entire product as one piece, which takes advantage of additive manufacturing's capabilities more fully. The design team switched to this approach and is moving forward with developing the part. Participants noted that retraining the manufacturing workforce to think differently about design will be an important part of building skills, including retraining experienced engineers who may be challenged by changes in traditional methods.

Designing for additive manufacturing also involves working across multiple disciplines to develop products. Several participants mentioned the importance of integrating skills from different disciplines into design for additive manufacturing, including science, technology, engineering, arts, and mathematics (STEAM). A workforce that can cross disciplinary boundaries, such as materials science, mechanical engineering, and manufacturing engineering, will be a key for additive manufacturing design, according to participants. For example, to develop products for additive manufacturing, a company may need a chemist who can relate to and work with materials specialists as well as be able to incorporate ideas from digital design into their work. In addition, some participants said that the kind of open-ended and creative thinking that is used in the arts will likely have value for the additive manufacturing workforce. For example, one participant said incorporating people with an arts background into design would be valuable because they think in fundamentally different ways than engineers and might use CAD software differently. Another participant noted that individuals from nonengineering backgrounds, such as graphic designers, often make very good designers for additive manufacturing.

Several participants said that the technology should be incorporated into educational and workforce training curricula because it can help prepare the future workforce. Following are examples:

- According to some participants, additive manufacturing should be incorporated into education from kindergarten through 12th grade to help engage students with the technology. This early engagement helps to build the pipeline of talent for the future workforce, according to participants.
- Several participants noted the importance of improving education at the university level. Participants said there are an increasing number of additive manufacturing degree programs, but some said a degree focused on additive manufacturing might not include the underlying skills needed to fully utilize the technology. One participant suggested that additive manufacturing should be incorporated into the existing curricula for traditional STEAM disciplines as well as into disciplines like medicine and law because the technology is likely to be extensively used across society.

- Participants also said that training in additive manufacturing could be a part of nondegree programs or 2-year technical degrees. For example, one participant said there is an opportunity for programs to train individuals to use additive manufacturing to fabricate items for medical professionals, such as models or guides for surgeons. Training individuals to use the technology for specific applications like this can be done in a matter of weeks as opposed to the more intensive training and education needed to learn to design for additive manufacturing, according to the participant.

The Additive Manufacturing Industry Faces Other Challenges to Supporting Increased Production of Functional Parts

Forum participants identified several other challenges the additive manufacturing industry faces in supporting increased production of functional parts. Specifically, participants agreed that the additive manufacturing industry needs a more robust industrial infrastructure, including stronger supply chains and better access to financing. The industry relies on sometimes limited supply chains for machines and materials, which can pose a challenge to the increased use of additive manufacturing for functional part production. For example, one participant mentioned an industrial accident at a facility that was the sole supplier of a nylon powder used in additive manufacturing machines. When that facility stopped producing material, there was nowhere else to get that powder, so the production of parts using that powder stopped. The robustness of a material supply chain depends on the material itself. For example, some metal-based additive manufacturing machines use metal welding wire, and there is a robust supply chain for that. However, more unique additive manufacturing materials, such as metal or nylon powders, typically have less robust supply chains.

Participants also expressed concerns about the robustness of the supply chain of additive manufacturing machines. One participant said that rapid growth of additive manufacturing has put pressure on the industry's supply chains to support that growth. As the industry continues to grow, there is a concern that the infrastructure may not be able to support that continued growth. For example, a participant mentioned an additive manufacturing company that is bidding on a major production contract to help the bidder transition from prototyping into direct functional part production. A company leader voiced concern that there may not be a reliable supply of additive manufacturing machines to meet the contract, meaning they would take a risk in bidding for the contract. Limitations on the supply of machines also can make it difficult for companies to scale up production, another challenge mentioned by participants, because producing larger volumes of parts can require many additive manufacturing machines. These supply chain challenges make using additive manufacturing to produce functional parts a potential risk for a project. These risks can hinder further adoption of additive manufacturing by managers who are focused on meeting schedule, cost, and performance requirements.

Participants noted challenges with obtaining financing for capital investments in additive manufacturing machines, which can be expensive.[19] One participant said the financing problem involves a misunderstanding of additive manufacturing compared with other manufacturing processes. For example, the participant said that, when considering an investment in a more established type of conventional manufacturing machine, he could calculate when that machine would start producing revenue based on how many parts it could produce in a given time period. However, with additive manufacturing machines, the fabrication speed—the time it takes for an additive manufacturing machine to build a given part—can be much slower than other manufacturing processes. As a result, the additive manufacturing machine may not appear to be a good investment when compared with an investment in a conventional manufacturing machine. However, this approach may not take into account some potential advantages of the additive manufacturing technology, such as the ability to manufacture small numbers of highly customized items such as prosthetics or its inherent flexibility. Several participants agreed that there needs to be a better understanding of the technology and its implications among financial stakeholders.

Several participants also highlighted the importance of understanding when additive manufacturing can be used cost-effectively. Participants agreed that additive manufacturing will not be the best process for every situation. To inform decisions about whether to use additive manufacturing, companies and other potential users need to analyze the costs and benefits of using the technology. Among things that would be helpful to consider, participants said, are cost models and life-cycle costs related to additive manufacturing. These decision-making tools can help potential users of the technology, such as project managers, better understand the risks and benefits of using additive manufacturing considering their schedules, costs, and performance goals. These decisions are particularly important for small- and medium-sized companies, which have more limited resources to take risks than large companies. Combined with other industry challenges, such as availability of materials and skilled workers, there are notable barriers to adopting additive manufacturing for small- and medium-sized companies, one participant said.

Participants identified the technical capabilities of additive manufacturing as challenges to supporting the technology's increased production of functional parts. For example,

- Several participants said that, compared with established manufacturing technologies, additive manufacturing has a limited set of materials that can be used to produce functional parts. In addition, the industry needs to design materials for additive manufacturing that are specially suited to new products. For example, one participant highlighted that additive manufacturing allows a greater level of customization in the medical device industry than previously available, including designing what the surface of a product should be like or how tissue can adhere to it. However, medical device companies looking to use additive manufacturing do not currently have the necessary materials to fully implement this potential level of customization.
- Another technical capability that is a challenge involves the ability to build larger and smaller products. For example, most additive manufacturing machines have a desktop-size building space, according to one participant, and that limits the ability to build large products.[20] On the other hand, improving the ability to do very small-scale manufacturing with the machines, such as laying down functional electronic circuitry, would be important.
- Participants identified fabrication speed as a challenge to increased production. For example, at current fabrication speeds, constructing a part can take hours or even days, compared with much shorter times for other manufacturing processes. That said, in select circumstances, additive manufacturing can reduce overall manufacturing times compared to conventional manufacturing.

Several participants commented that improving the technical capabilities of additive manufacturing can be accomplished with more research and development, much of which is under way.

Participants identified challenges related to coordination and information sharing among those involved with additive manufacturing technology. For example, participants generally agreed that more research and development will be needed to advance additive manufacturing and increase its use in making functional parts. However, they identified challenges related to coordination of research and development necessary to make these advances. For example, some participants expressed concern over uncoordinated duplication of research efforts, such as redundant efforts to create material properties databases. However, other participants pointed out that duplication of research can have value at this stage in additive manufacturing's technical development if it is used to validate previous findings, a key step in the scientific process. While participants generally agreed that more research and development is key to addressing the technical challenges of additive manufacturing, several expressed concerns that the industry will have difficulty coordinating this research and development because research findings can lead to a competitive advantage in the marketplace. Research on additive manufacturing, such as collecting data on material properties, represents potentially valuable proprietary information for companies that they may not want to share. Several participants agreed that identifying areas where information could be shared and not lead to a competitive advantage (also called precompetitive research) would be beneficial to the industry, but that doing so would be difficult and requires substantial coordination.

Key Considerations for Federal Policy Actions That Could Affect the Future Use of Additive Manufacturing for Production of Functional Parts

Forum participants identified three key areas that federal policymakers would need to consider before taking any policy actions that could affect the future use of additive manufacturing for production of functional parts: (1) key challenges facing the industry, (2) areas that could be affected by an increased use of additive manufacturing, and (3) trade-offs of any potential actions. There was no consensus on specific policy actions needed, but participants discussed a variety of policy approaches that the government could consider, including encouraging a national dialogue about the government's role.

Participants Highlighted Key Industry Challenges That Could Benefit from Government Involvement

As we discussed previously, the additive manufacturing industry faces several challenges to support increased production of functional parts and forum participants stated that all of these challenges could benefit from increased government involvement. The three key areas identified by participants as the most important for federal involvement are: (1) standards setting, (2) workforce and training, and (3) market support and readiness.

Forum participants said that if federal actions focused on standards setting, they could encourage increased use of additive manufacturing. Participants discussed the federal government's roles as both an entity that develops and applies measurements and standards and as a consumer of additively manufacturing goods with specific requirements. One forum participant emphasized the importance of establishing standards by describing them as the ground rules for opportunities in the sector. The federal government, primarily through NIST, currently plays a role in setting standards and requirements for additive manufacturing in collaboration with industry-led standards developing organizations, such as ASTM International. However, one participant said that the process is slow, and the work is generally done by a volunteer workforce from government, industry, and academia. Participants suggested that the government could be a catalyst to provide momentum to get more participants involved in standards setting. In some cases, not enough is known about the processes and materials used in additive manufacturing to establish standards, so NIST conducts research in four key areas: (1) characterizing materials and mechanical properties of manufactured parts, (2) creating standard methods for measuring and improving process repeatability and accuracy, (3) developing fundamental data and methods for qualification of processes and parts, and (4) conducting system analysis of the entire additive manufacturing information flow from beginning to end.

According to forum participants, government consumers, such as DOD and NASA, have specialized missions with unique product reliability requirements. Both agencies are interested in the potential use of additive manufacturing, but using additively manufactured products in their missions will be challenging because they require assurances of both the consistency and quality of any products they use. A forum participant stated that DOD has an ongoing project to develop products using additive manufacturing, which has required a redesign of some of its process steps to accommodate its needs for quality inspection using nondestructive methods. Participants also cited other areas of ongoing work at regulatory agencies, such as the Federal Aviation Administration and the Food and Drug Administration, to ensure that additively manufactured goods are capable of meeting their quality and performance requirements.

In addition to standards setting, forum participants emphasized the importance of federal actions in the area of workforce education and training. Participants frequently mentioned the need for engineering and technical education programs to increase their focus on design education. Additive manufacturing fundamentally restructures the possibilities of manufacturing, and forum participants said that a workforce capable of design-focused thinking is important to additive manufacturing's success. Federal agencies currently play a significant role in developing the capabilities of the U.S. workforce through numerous programs to support science, technology, engineering, and mathematics education. A forum participant noted that a recent GAO report identified over 200 federally-funded science, technology, engineering, and mathematics programs.[21] Federal agencies also have programs that support education in the arts and design. Forum participants noted that adding design thinking into more traditional disciplines could be beneficial to the sector. One participant described the current education and workforce development situation as the "Wild West" and discussed the need for an education organization and coordination plan for the United States with a goal of improving the situation. Another participant suggested that other organizations along the public-private partnership spectrum might be able to help coordinate education and workforce issues. Another forum participant discussed the need for directed workforce training efforts and suggested a role for community colleges and vocational/technical schools.

Forum participants identified additive manufacturing market support and readiness as the third key area of federal action needed to support the production of functional parts in the United States. A participant suggested looking at the federal acquisition strategy and determining where it would be possible and practical to make spare parts by additive manufacturing. A participant described this as an investment in the future of the technology in the federal acquisition system. In addition to its function as a consumer, the federal government works to integrate U.S. companies with existing public, private, or not-for-profit industrial and economic development resources. Currently, this is done in partnership with America Makes, which has a stated purpose of fostering collaborative infrastructure for the additive manufacturing industry. However, several participants expressed concern that collaboration between private and university researchers may not take place due to competition issues, and that the industry as a whole might suffer. Another participant suggested that investments in additive manufacturing made by one company would challenge the rest of the industry to keep pace. The participant said that, even though a company might not share its proprietary information, it might encourage others by its progress. The same participant also identified the development of mathematical analytical and design tools, as well as training efforts, as fundamental work that the government could support to move the industry forward, regardless of information sharing practices between companies. Another participant added that the federal government could encourage the purchase and use of additive manufacturing equipment with a tax subsidy, such as a bonus depreciation tax deduction.[22]

Participants Said the Government Should Monitor Areas That Could Be Affected by an Increased Use of Additive Manufacturing

Forum participants identified four key areas that could be affected by future growth in additive manufacturing: (1) national security, (2) product liability, (3) intellectual property, and (4) environmental, health, and safety concerns. These four areas were identified as being particularly challenging and perhaps different from challenges caused by other emerging technologies. Most participants agreed with the call for a "wait and see" posture with the government closely following developments in the field and ready to act when required, but no sooner.

Forum participants said that developing and maintaining an infrastructure that supports widespread use of additive manufacturing technologies may raise national security issues. Specifically, they raised the need to balance the national security benefits of a strong U.S. industrial base with the national security risks related to increased use of additive manufacturing. For example, they said hostile nations or factions could use the technology to harm the United States or its representatives abroad, and the military and intelligence communities are particularly concerned with the potential for additive manufacturing to permit the production of weapons or weapons components in ways that are less observable than with current conventional manufacturing processes and facilities. For instance, additive manufacturing could supplement the proliferation of small arms and light weapons. Because additive manufacturing has been recognized as a technology with important military applications by DOD, finding the appropriate mechanisms to avoid damaging the industry, while protecting national security, will likely be an ongoing challenge. Many participants expressed concern that overly restrictive export controls or licensing requirements on additive manufacturing technologies could be an impediment to the economic growth of U.S. companies. One participant said that the rationale for restrictive export controls presupposed a U.S. monopoly on the technology, which the participant asserted was not the case. Another participant said there is a need to balance a protectionist, "Fortress America" view of national security with the benefits of positively engaging with global commerce. One participant also emphasized that the cybersecurity capabilities of a system need to be built-in and considered in the design and development of the product, including design files, materials, and machines.

Participants cited uncertainty about product liability protections when additive manufacturing technology is used by individual consumers with home 3D printing systems. A forum participant said this development could complicate traditional regulatory roles and result in insufficient legal protections for those harmed by dangerous products. There is also a challenge in regulating the safety of final products produced through additive manufacturing by small organizations and individuals. These small organizations and individuals are less likely to be aware of federal safety regulations for final products and may not have the quality assurance systems that larger organizations do. For example, individual consumers injured by a product printed with a home 3D printing system may have difficulty seeking legal recourse because strict product liability applies only to commercial sellers.[23] However, it is unlikely the individual consumer will have a commercial seller to sue for the defective product, especially if the consumer uses a 3D printer to manufacture the product at home. Therefore, it is uncertain who, if anyone, could be held liable for a defective product manufactured with a home 3D printer. One participant noted that liability law historically has adapted to new situations as they have emerged, but that policymakers should be prepared to make changes if problems arise.

Forum participants also discussed the potential risk for infringement of intellectual property rights with regard to additive manufacturing equipment, design files, and product manufacturers. They said it is possible that an increased use of additive manufacturing and digital design files could lead to disputes between additive manufacturing users and intellectual property holders. Currently, a limited number of products can be effectively produced by additive manufacturing, particularly on the machines available at a low cost to the consumer. However, as the technology improves and proliferates, machines will likely be capable of building more complex items and will become more available to consumers. Moreover, 3D scanning technology and sharing of design files on the Internet may make it easier to infringe upon intellectual property rights. Participants noted that, while copying and counterfeiting exist in every industry, 3D scanning and additive manufacturing technologies may make such activities easier.

Finally, participants discussed whether, and to what extent, the government should regulate environmental, health, and safety (EHS) concerns when implementing and using additive manufacturing processes. One participant noted that some EHS concerns for additive manufacturing are not markedly different from conventional manufacturing techniques such as welding, but some other concerns are unknown. EHS concerns unique to additive manufacturing have not been clearly identified, but it may be that safety practices for handling materials and managing equipment could be similar for additive and conventional manufacturing. Forum participants also raised EHS concerns related to the maker community's use of 3D printing. In a large manufacturing organization, safety assessors monitor new machines and set-ups and determine the appropriate personnel and environmental safeguards. The participants expressed some concern that novice or home users would not be able to recognize or resolve EHS concerns associated with additive manufacturing.

Participants Encouraged Government Consideration of Trade-offs for Potential Policy Actions Affecting Additive Manufacturin

Forum participants said that as the government considers policy actions to address challenges that could advance additive manufacturing, as well as the possible risks associated with an increased growth in the sector, it needs to consider the trade-offs involved in such actions. Trade-offs exist for any government action and participants encouraged government decision makers to acknowledge that there is a cost to government intervention. They suggested it would be worth waiting to see what is possible and what the real state of the art is before, as one participant put it, "running off to try to intercept a concern." Another participant described well-exercised government control for additive manufacturing as the crafting of solutions to only those problems and concerns that can be clearly articulated.

Given the constraints on federal resources, participants questioned how deeply the federal government should be involved in promoting specific applications of additive manufacturing. For instance, as previously mentioned, while there is a lot of attention being paid to additive manufacturing, participants said that additive manufacturing will not replace conventional manufacturing, but it will be an additional tool for manufacturing parts to be used, when appropriate, from a cost-benefit perspective. Further, forum participants discussed whether additive manufacturing-specific training should come from government efforts, or from industry, and to what degree additive manufacturing technical details might be taught on-the-job to a workforce with sufficient training in design. Concerning market support, participants discussed whether or how the government should take steps to subsidize the early stages of technology development or provide incentives to program managers to mitigate the real or perceived risk of taking on additive manufacturing projects. Forum participants suggested that the federal government could support new technologies through tax or other revenue policies.

Participants encouraged a national dialogue about government involvement in additive manufacturing for production of functional parts, which could include a substantive discussion of the trade-offs inherent in government involvement. Participants said that clearly articulated goals for additive manufacturing would help policymakers assess trade-offs associated with government action, and react accordingly. One forum participant noted that the development of such high-level goals must come from a consensus process, and they cannot be developed by the government alone. For example, forum participants said one goal for additive manufacturing could be enhancing the international competitiveness of the United States. Participants noted that additive manufacturing was originally developed in the United States, but the production of additive manufacturing machines and materials now is occurring in countries throughout the world, especially in Sweden, Germany, and China. One participant stated that, when a country does not control the technology and the manufacturing of the equipment itself, it becomes difficult to be a leader in the use of that equipment. Participants suggested that enhancing the ability of U.S. companies to compete in the global marketplace could be assisted by America Makes, or a similar public-private partnership. The administration has voiced support for advanced manufacturing, including the ability to quickly respond to customer needs through innovations in production processes and the supply chain. Participants suggested that the United States focus discussions on where the federal government can enable U.S. strengths and level the playing field with other nations that are making investments in additive manufacturing capabilities.

The discussion among forum participants also generated a list of possible focus areas: strategic planning and idea generation for the sector, ways to ramp up production of additive manufacturing machines, deployment of the technology, the creation of analytical or design software tools, and the development of new materials. The exact mechanisms to be used in pursuing any of these focus areas were not identified. Instead, participants emphasized the need for broad thinking to select and design the most effective initiatives. Participants said that collaboration, coordination, and communication were critical areas in which the federal government could help build the body of knowledge and take other steps that could move the technology forward.

APPENDIX I: OBJECTIVES, SCOPE, AND METHODOLOGY

This report summarizes discussions from the Comptroller General Forum on Additive Manufacturing that was held on October 15 and 16, 2014. The forum focused on three questions: (1) What are the opportunities to produce functional parts using additive manufacturing? (2) What are the key challenges, if any, to using additive manufacturing to produce functional parts? and (3) What are the key considerations for any policy actions that could affect the future use of additive manufacturing to produce functional parts? To inform our understanding of additive manufacturing and to establish the scope and agenda for the forum, we met with relevant agency officials from the National Institute of Standards and Technology, the U.S. Patent and Trademark Office, the U.S. Copyright Office, and the Consumer Product Safety Commission. We also met with key additive manufacturing stakeholders, such as the National Additive Manufacturing Innovation Institute (America Makes), the Association for Manufacturing Technology, Stratasys, and EWI.[1] We attended additive manufacturing conferences sponsored by Oak Ridge National Laboratory and by SME.[2] We also conducted a comprehensive literature review. We reviewed literature on technical and policy issues related to additive manufacturing. Articles selected for the literature search were found in one of the following three ways:

1) searches of databases through RefWorks,
2) interviews with relevant agency officials and other key informants (i.e., individuals identified as having relevant expertise or knowledge

about technical and policy issues related to additive manufacturing), and

3) references in literature.

To determine whether the selected articles were relevant, we reviewed each one and excluded those that were not related to the use of additive manufacturing for the production of functional parts. For instance, we excluded articles on rapid prototyping, home-user-produced items, particular additive manufacturing printer technologies, or niche applications. We focused on broad technical and policy issues to address the researchable questions. When an article was excluded, we noted the reason for exclusion. In total, nearly 160 documents were analyzed as part of this literature review. Each of the relevant documents was reviewed by a primary and secondary analyst, and the following information was recorded, as appropriate:

1) ID – the Ref ID from the RefWorks search results;
2) title – the title of the article;
3) source – the title of the journal where the article appeared or other source of the article;
4) publication year – the year that the article was published;
5) authors/contact – the author(s) or responsible contact(s) provided in the article;
6) company/agency/industry use of additive manufacturing objects – companies, agencies, or industries identified as using additive manufacturing objects and any provided information about what objects they are using and how they are using them;
7) technical and policy challenges and opportunities identified – any technical and policy issues discussed in the article, including limitations, risks, benefits, and other important issues related to the advancement and future of additive manufacturing; and
8) other potential sources identified – any other potential sources (e.g. publications, experts, etc) that may provide information about how additive manufacturing objects are being used and technical and policy issues related to additive manufacturing.

We used the results of the literature review, as well as information obtained from interviews with experts, agency officials, and other key informants and gathered through attending relevant conferences, to identify and categorize the technological and policy issues. We also used the information to determine the various sessions that comprised the forum agenda. We prepared background materials and discussion papers that framed the issues to be discussed during each session. The identification and categorization of technological and policy issues also aided in determining the needed expertise at the forum. To identify participants for the forum, we contracted with the National Academies to assist us.[3] Overall, the participants included individuals with a range of backgrounds, experiences, and knowledge who comprised the following two main groups:

- Representatives from government and business involved with current or planned uses of additive manufacturing to produce operational parts and products who could address current and potential applications, benefits, and challenges.
- Experts who could address how the use of additive manufacturing may affect areas such as supply chain management, product safety, intellectual property protection, and national security, as well as potential changes to existing policy frameworks to address additive manufacturing challenges.

To identify experts for the forum, the National Academies staff received input and recommendations from several of its internal boards and advisory groups, including the National Materials and Manufacturing Board and its standing committee on Defense Material Manufacturing and Infrastructure; Space Studies Board; Aeronautics and Space Engineering Board; Board on Physics and Astronomy; Laboratory Assessments Board; and the Computer Science and Telecommunications Board. The staff then acquired biographical information on the suggested experts and compared their backgrounds with the topics for the forum. The National Academies staff made initial contact with potential experts to determine their interest and availability for the forum. The National Academies presented GAO a preliminary list of experts for the meeting. GAO provided feedback on the list, and the National Academies staff made revisions to the list. From the revised list, GAO selected 23 meeting attendees, which were subsequently invited by the National Academies.[4] GAO made its selections based on the candidates' biographies to ensure a suitable number of experts to cover all the necessary topics to be covered in the forum and to ensure balance and representation on a wide range of viewpoints. From each of these attendees, the National Academies collected a questionnaire to identify any potential conflicts of interest, including investments and other assets, sources of earned income, organizational positions, and any other relevant circumstances.

The Comptroller General Forum on Additive Manufacturing was held at the GAO headquarters building in Washington, D.C. The forum was recorded, and the sessions were transcribed. In total, 22 experts, as well as several senior level GAO staff participated in the forum. The forum participants are listed in appendix II, and the agenda is included in appendix III. All sessions were structured as facilitated discussions.

The focus of the first day was to discuss the opportunities for additive manufacturing to produce functional parts and to identify the technical and policy challenges of using additive manufacturing in this way. At the conclusion of the first day of the forum, we compiled a list of 35 challenges that were raised during the sessions that are limiting the use of additive manufacturing for producing functional parts. This list of challenges was also informed by the information obtained in advance of the forum. At the beginning of the second day of the forum, we presented this list of challenges to the participants. To ensure the accuracy of the list, we asked the participants to indicate whether or not they agreed that we had accurately characterized the challenges, whether anything was missing, and whether any challenges could be combined. We edited the list in response to the participants' feedback. Following the review, we had the list of 37 challenges that are included in appendix IV.

To determine the participants' views on the most important challenges, we prepared a document listing all the challenges and distributed it to each participant. The participants were instructed to choose the 10 challenges they considered most important, in no particular order. The participants' selections were tabulated and presented for further discussion later in the afternoon. The original intent was to include a second round of challenge selection to narrow the list to the top three most important challenges. However, participants decided instead to group the challenges into categories they considered most important.

Following the forum, to provide a framework to present the additive manufacturing challenges in this report, we allocated specific challenges identified by forum participants into broad categories.

At the beginning of the analysis, we developed three broad categories for various challenges identified by the participants. The initial set of categories was based generally on challenges related to different stages of the additive manufacturing process (design, construction, and verification). Specifically, we developed the following questions based on those categories:

1) Challenges to envisioning use of the technology – can we design what we want to build?
2) Challenges to using the technology – can we build what we designed?
3) Challenges to verification – can we verify we built what we intended to build?

We determined the challenges most often identified as among the 10 most important challenges, and we ended with challenges that were the least often identified as among the top 10. Multiple GAO analysts were involved and any disagreements were worked through to achieve consensus. For example, the challenge "Consensus methods and test data for qualification and certification are not sufficient" was placed in Category 3 (challenges to verification) because qualification and certification relate to verifying that the product and its materials meet a set of standards or performance criteria.

We also decided many challenges were related and could potentially be classified into multiple categories. When a challenge was classified in more than one category, all relevant category numbers were listed in order from most relevant category to least relevant. For example, the challenge "Insufficiently skilled workforce and training" was actually placed in all of the categories because skilled workers will be necessary for every aspect of the technology. However, we selected Challenge 1 (challenges to envisioning use of the technology) as the first challenge because participants repeatedly emphasized the need for a new approach to thinking about design and the underlying education needs.

We determined that the forum included discussion of challenges that currently faced the additive manufacturing industry, as well as challenges that might potentially face the industry and the government as the technology continues to develop. The determination of whether a challenge was primarily a current challenge or a potential challenge became the first level of categorization. For example, the challenge "Need for appropriate design and mathematical analysis tools for additive manufacturing products" is categorized as a current challenge because this is a problem that the industry currently faces and limits its use of the technology. However, the challenge "Difficulty in regulating final additive manufacturing products when made by laypersons" is classified as a potential challenge because there are potential problems in regulating distributed manufacturing, but these problems have not actually manifested according to participants. When a challenge was classified as potential, we did not categorize it into any numbered category.

We identified another theme in our discussion of the challenges and added a fourth category labeled "Challenges related to knowledge gaps and the need for greater understanding of technology/implications." This theme came up in forum discussions and was added to our list to better capture challenges that involved uncertainty or the need for better coordination/collaboration to build a body of knowledge. These challenges tended to involve other categories as well.

Shortly after the forum, we prepared an outline of this report and distributed it to the forum participants for their comments. Nineteen of the 22 experts who participated in the forum responded to our request for comments on the outline of the report.[5] Participants provided several technical comments, but they generally agreed with the framing of the challenges. As we drafted the report, we incorporated participants' comments, as appropriate. In addition, we provided a draft of this report to the forum participants for their comments. Eighteen of the 22 experts who participated in the forum responded to our request for comments on the draft report.[6] Participants provided several technical comments, but they generally agreed that the report accurately reflects the forum discussions. We incorporated participants' comments into the final report, as appropriate.

We conducted our work from March 2014 to June 2015 in accordance with all sections of GAO's Quality Assurance Framework that are relevant to our objectives. The framework requires that we plan and perform the engagement to obtain sufficient and appropriate evidence to meet our stated objectives and to discuss any limitations in our work. We believe that the information and data obtained, and the analysis conducted, provide a reasonable basis for any findings and conclusions in this product.

APPENDIX II: FORUM PARTICIPANTS

The Honorable Joseph Benkert
Former Assistant Secretary of Defense for Global Security Affairs
Vice President
The Cohen Group

Mr. Anthony DeCarmine
Chief Technology Officer
Oxford Performance Materials

Mr. Carl Dekker
President
Met-L-Flo Inc

Dr. Phill Dickens
Professor of Manufacturing Technology
University of Nottingham

The Honorable Gene L. Dodaro
Comptroller General of the United States
U.S. Government Accountability Office

Mr. Michael Dunn
Manufacturing Technology (ManTech) Program
Department of Defense

Ms. Nora Engstrom
Professor
Stanford Law School

Dr. Slade Gardner
Fellow
Lockheed Martin Corporation

Capt. Gerald Grant
Service Chief, 3D Medical Applications Center
Director, Craniofacial Imaging Research
Walter Reed National Military Medical Center

Mr. Matthew K. Hlavin
Chief Executive Officer
Rapid Prototype and Manufacturing LLC

Mr. John Hornick
Partner
Finnegan, Henderson, Farabow, Garrett & Dunner, LLP

Mr. Kevin Jurrens
Deputy Chief, Intelligent Systems Division
National Institute of Standards and Technology

Dr. Nancy Kingsbury
Managing Director, Applied Research and Methods
U.S. Government Accountability Office

Dr. Howard A. Kuhn
Adjunct Professor
Swanson School of Engineering
University of Pittsburgh

Dr. Thomas Kurfess
Professor
HUSCO/Ramirez Distinguished Chair, Fluid Power and Motion Control
Georgia Institute of Technology

Dr. Aija Leiponen
Associate Professor
Dyson School of Applied Economics and Management
Cornell University

The Honorable Donald Manzullo
Former United States Congressman representing
the 16th District of Illinois

Dr. Michael McGrath
Vice President
Analytic Services Inc.

Mr. Ed Morris
Vice President and Director
National Center for Defense Manufacturing and Machining
America Makes – The National Additive Manufacturing
Innovation Institute

Dr. Susan Offutt
Chief Economist
U.S. Government Accountability Office

Mr. Neal Orringer
Vice President of Alliances and Partnerships
3D Systems

Mr. Michael Peretti
Director, Materials & Manufacturing Advanced Programs
General Electric Aviation

Dr. Timothy M. Persons
Chief Scientist
U.S. Government Accountability Office

Ms. Susan A. Poling
General Counsel
U.S. Government Accountability Office

Dr. Steve Pollack
Director, Office of Science and Engineering Laboratories
Food and Drug Administration

Ms. Michelle L. Proctor
Director of Innovation
FedEx

Mr. Michael Weinberg
Vice President
Public Knowledge

APPENDIX III: FORUM AGENDA

Comptroller General Forum on Additive Manufacturing
October 15-16, 2014

Agenda

Wednesday October 15, 2014		
8:00 a.m.	**Arrivals/Check-in** Continental Breakfast	McCarl Room
8:30 a.m.	**Opening Session**	Staats Briefing Room
	Welcome and Introductions • The Honorable Gene L. Dodaro Comptroller General of the United States U.S. Government Accountability Office *Forum Overview* • Dr. Timothy M. Persons, Chief Scientist U.S. Government Accountability Office *Significance of Additive Manufacturing to the Congress* • The Honorable Evan Bayh, former United States Senator and former Governor of Indiana • The Honorable Donald Manzullo, former United States Congressman	
9:15 a.m.	**Opportunities for Additive Manufacturing to Directly Produce Parts and Products**	Staats Briefing Room
11:00 a.m.	**Break**	
11:15 a.m.	**Technical Challenges of Additive Manufacturing** Working Lunch	Staats Briefing Room McCarl Room (lunch pickup)
1:30 p.m.	**Issues Affecting the Development and Adoption of Additive Manufacturing**	Staats Briefing Room
3:30 p.m.	**Break**	
3:45 p.m.	**Policy Implications Arising From Using Additive**	Staats Briefing Room

(Continued)

	Manufacturing for Direct Part Production	
5:30 p.m.	**Reception**	McCarl Room
Thursday, October 16, 2014		
8:00 a.m.	**Arrivals**	McCarl Room
	Continental Breakfast	
8:30 a.m.	**Review of Issues Discussed on Day 1**	Staats Briefing Room
9:00 a.m.	**Importance of Addressing Additive Manufacturing**	Staats Briefing Room
	Issues for Direct Part Production	
10:30 a.m.	**Break**	
10:45 a.m.	**Key Considerations for Government Policies to Address Additive Manufacturing Issues for Direct Part Production** Working Lunch	Staats Briefing Room McCarl Room (lunch pickup)
1:30 p.m.	**Forum Summary and Concluding Remarks**	Staats Briefing Room
2:00 p.m.	**Adjourn**	

Appendix IV: Challenges Identified by Forum Participants

Forum participants identified the following challenges, listed in no particular order, that potentially limit the use of additive manufacturing for functional part production:

- Consensus methods and test data for qualification and certification are not sufficient
- Need for improved suite of standards for additive manufacturing
- Lack of inspection and quality standards for additive manufacturing products
- Need for better process control to improve system performance and repeatability in additive manufacturing
- Need for appropriate design and mathematical analysis tools for additive manufacturing products

- Need for appropriate design and physics-based analysis tools for additive manufacturing processes
- Limited number of additive manufacturing materials
- Need to print smaller-scale and larger-scale parts
- Difficult to scale up production
- Improved part accuracy
- Improved surface finish
- Unknown materials properties including materials compatibility
- Fabrication speed too slow
- Need to make more complex or integrated parts and products and not just shapes using additive manufacturing
- Cybersecurity risks to additive manufacturing equipment and design files
- Insufficient sharing of common data on additive manufacturing process parameters, materials, materials properties, and testing
- Need for improved collaboration and coordination among researchers
- Need for more pragmatic output from university research and development
- Need for more additive manufacturing research and development infrastructure (labs, lab equipment)
- Hype and unrealistic expectations (e.g., disconnect between expectations and reality)
- Technology's reputation may be tarnished by bad experiences or by association with criminal activity
- Insufficiently skilled workforce and training
- Uncertainty regarding where additive manufacturing is cost-effective
- Pace of technology change may result in relatively short life-cycle or obsolescence of the equipment (5-10 years)
- Effect of additive manufacturing equipment or software revisions on process qualification
- Need for funding or access to capital for additive manufacturing investments
- Need for robust supply chain for additive manufacturing materials and systems
- Lack of incentives for program managers to take on perceived or actual risk associated with additive manufacturing
- Need to identify breadth and depth of precompetitive discussions for additive manufacturing

- Potential clash between additive manufacturing and intellectual property stakeholders
- Environmental health and safety concerns when implementing and using additive manufacturing process
- Difficulty in regulating final additive manufacturing products when made by novices
- Potential safety concerns and uncertainty of product liability protections when additive manufacturing is used by novices
- Need to avoid export controls/licensing that are impediments to U.S. additive manufacturing industry growth while still protecting U.S. national security
- Need for greater consideration of U.S. industrial base and national security concerns for additive manufacturing
- Need for improved coordination among federal agencies on additive manufacturing issues
- Lack of national strategy to enhance U.S. competitiveness and engagement in the global marketplace

End Notes

[1] Additive manufacturing and 3D printing are terms that are often used interchangeably, as is the case in this report.

[2] "3D Printing: The Printed World," *The Economist* (Feb. 10, 2011).

[3] National Defense University, "Toward the Printed World: Additive Manufacturing and Implications for National Security," *Defense Horizons* (September 2012).

[4] Gartner, *Hype Cycle for 3D Printing, 2014* (July 21, 2014).

[5] National Institute of Standards and Technology, *Economics of the U.S. Additive Manufacturing Industry*, NIST Special Publication 1163 (August 2013).

[6] ASTM International, a standards development organization, was formerly known as the American Society for Testing and Materials.

[7] Science and Technology Policy Institute, *The Role of the National Science Foundation in the Origin and Evolution of Additive Manufacturing in the United States*, IDA Paper P-5091 (November 2013).

[8] Bioresorbable materials can dissolve into the body.

[9] A braze is a soldered joint between metal pieces.

[10] A treatment plan describes a patient's condition and necessary procedures, including details of the treatment to be provided, the expected outcome, and the expected duration of the treatment.

[11] GAO, *Defense Inventory: Management Actions Needed to Improve the Cost Efficiency of the Navy's Spare Parts Inventory,* GAO-09-103 (Washington, D.C.: Dec. 12, 2008).

[12] Standards include common and repeated use of rules, conditions, guidelines, or characteristics for products or related processes and production methods, and related management systems

practices. As of March 2015, ASTM International had 11 standards for additive manufacturing in design (1), materials and processes (6), terminology (1), and test methods (3).

[13] For example, DOD requires parts it purchases, such as aircraft engine parts, to meet specific standards or performance criteria. Manufacturers might need to have these parts certified to meet DOD's standards.

[14] For example, the Federal Aviation Administration sets standards in its regulations for safety critical airplane parts, such as engine parts. Manufacturers must certify that any safety critical parts meet these standards by having them certified.

[15] Consensus methods are standard testing methods that could be applied across the additive manufacturing industry and used in the qualification and certification process.

[16] Qualification processes are used to determine whether a product, process, or system meets a set of established requirements. Materials and processes used to produce critical components for applications in the defense, aerospace, and medical industries must be formally qualified before the components can be certified. Extensive empirical testing to fully qualify a material often requires conducting thousands of tests, costing millions of dollars, and taking years to complete.

The destructive testing approach involves testing a statistically representative sample of parts to determine whether they meet product requirements as well as for reliability and durability. For example, destructive testing is often used to determine the physical properties of materials, such as impact resistance. This testing is more commonly used in high-volume production where the loss of a set number of parts for testing is not a substantial cost.

[18] Specialized tools, such as touch probes, are designed to measure component surfaces, providing accurate, geometric data. Obtaining and analyzing these data throughout the manufacturing process can help to ensure components remain within conformance limits.

[19] Additive manufacturing machines for industrial use can range in cost from hundreds of thousands of dollars to a million dollars, or more.

[20] Participants said that machines with larger capacity exist, and that even larger ones are under development. However, most of the available machines have a relatively limited building space.

[21] GAO, *Science, Technology, Engineering, and Mathematics Education: Strategic Planning Needed to Better Manage Overlapping Programs across Multiple Agencies*, GAO-12-108 (Washington, D.C.: Jan. 20, 2012).

[22] Taxpayers are allowed to take an income tax deduction to cover the depreciation of certain property used in a business. A bonus depreciation allows the taxpayer to deduct more of the costs of the investment in the first year.

[23] Persons engaged in the business of selling or distributing products may be liable for harm to others caused by a defect in the product sold. Strict product liability holds those sellers or distributers legally responsible for the harm caused by the defective product, even if the seller did not intentionally or negligently cause the defect.

End Notes for Appendix I

[1] EWI was formerly known as the Edison Welding Institute.

[2] SME was formerly known as the Society of Manufacturing Engineers.

[3] GAO has a standing contract with the National Academies under which the National Academies provides assistance in convening groups of experts to provide information and expertise in our engagements.

[4] Though they were initially available, two of the invited experts were unable to attend the meeting. One of the invited experts designated a replacement to attend the forum meeting.

[5] Three forum participants did not respond to our request for comments on the report outline.

[6] Four forum participants did not respond to our request for comments on the draft report.

In: Additive Manufacturing
Editor: William H. Phillips

ISBN: 978-1-63484-232-7

Chapter 2

ADDITIVE MANUFACTURING TECHNOLOGY: POTENTIAL IMPLICATIONS FOR U.S. MANUFACTURING COMPETITIVENESS*

Sharon Ford[1]

ABSTRACT

This article explores the development and application of additive manufacturing as well as initiatives in the United States and other countries to advance it. It also examines the technology's effect on firm and industry production activities, as well as the potential implications for U.S. manufacturing competitiveness focused in three industries. It concludes that the most significant factors affecting the potential of additive manufacturing to contribute to U.S. competitiveness are developing standards, improving the selection and affordability of materials, and increasing the accuracy and reliability of equipment and processes.

---

* This is an edited, reformatted and augmented version of an article that originally appeared (online) in the United States International Trade Commission's Journal of International Commerce and Economics, September 2014.

INTRODUCTION

Additive manufacturing, or three-dimensional (3-D) printing, is receiving unprecedented attention. Additive manufacturing is a suite of emerging technologies that fabricates three-dimensional objects directly from digital models through an additive process, typically by depositing and "curing in place" successive layers of polymers, ceramics, or metals.[2] Unlike traditional manufacturing processes involving subtraction (e.g., cutting and shearing) and forming (e.g., stamping, bending, and molding), additive manufacturing joins materials together to build products. The number of articles published on this nascent industry rose from about 1,600 in 2011 to 16,000 in 2012.[3] The additive manufacturing market, consisting of all additively manufactured products and services worldwide, shows equally impressive growth: it rose from $1.7 billion in 2011 to $2.2 billion in 2012, by 28.6 percent.[4] Evolving and fluid, additive manufacturing technology is shaping the future of product development and manufacturing.

This article addresses three questions about additive manufacturing. First, how is this rapidly evolving technology being applied? Second, will its growth prompt changes in firm and industry production activities? Third, what are the potential implications for U.S. competitiveness in manufacturing processes and the economy? The article draws on the experiences of three industries that are thus far among the top users of additive manufacturing: automotive, medical, and aerospace.[5] It examines how additive manufacturing is used in these industries. The article also reviews the dynamics that affect wider deployment of additive manufacturing, such as technological challenges, new innovations, and industry and government initiatives to facilitate its use. It concludes that the most significant factors affecting the potential of additive manufacturing to contribute to U.S. competitiveness are developing standards, improving the selection and affordability of input materials, and increasing the accuracy and reliability of equipment and processes.[6]

Originally conceived as a way to make prototypes,[7] additive manufacturing has improved to the extent that it is increasingly used to deliver final products.[8] Recent improvements include enhancements of the speed and performance of additive manufacturing machinery, an expanding range of input materials, and falling prices for both machinery and materials. These advancements are likely to inspire further adoption of additive manufacturing in the United States and around the world in coming years.

Additive manufacturing provides an important opportunity to advance the U.S. manufacturing industry, which has the largest research and development

(R&D) expenditure for manufacturing overall of any country.[9] Though barriers to production with the technology exist, unique capabilities make additive manufacturing processes superior to conventional manufacturing for some products. These capabilities include constructing previously impossible geometries, such as pyramidal lattice truss structures with hollow trusses,[10] and significantly reducing material waste by building layer by layer and using only the material necessary.[11] Firms that employ additive manufacturing are beginning to achieve benefits such as increasing supply chain efficiencies; reducing time to market; moving from mass production to mass customization; and sustaining the environment.[12] As a result, the technology is receiving attention in policy as well as manufacturing circles. President Obama extolled additive manufacturing in his 2012 State of the Union address, stating it could "revolutionize the way we make almost everything."[13]

Overview of Technology

Process

Additive manufacturing begins with computer-aided design (CAD) modeling software that takes a series of digital images of a design or object and sends descriptions of them to a professional-grade industrial machine. The machine uses the descriptions as blueprints to create the item by adding material layer-upon-layer. Layers, which are measured in microns, are added by the hundreds or thousands until a three-dimensional object emerges. Raw materials may be in the form of a liquid, powder, or sheet and are typically plastics and other polymers,[14] metals, or ceramics.

The numerous additive manufacturing processes differ according to the materials and methods of patterning and fusing layers they employ. Major processes include material extrusion,[15] material jetting,[16] binder jetting,[17] sheet lamination[18] vat photopolymerization,[19] powder bed fusion,[20] and directed energy deposition.[21] Some of these melt or soften material to produce the layers, while others cure liquid materials using different sophisticated technologies.[22] An image of selective laser sintering, an additive manufacturing process that is a type of powder bed fusion, appears below (box 1) as an example.

Thirty-one manufacturers from around the world produced professional-grade industrial additive manufacturing machines in 2011, compared to 32 in 2010 and 35 in 2009. In 2010 and 2011, 9 of these companies sold more than

100 machines each. Firms that produce additive manufacturing machines range from those that produce and sell fewer than 10 per year, to those that sell hundreds of machines per year.[23]

The United States leads in the production and sales of professional-grade industrial additive manufacturing machines, with 35,753 units sold between 1998 and 2011. Israel and Germany made 4,556 and 3,980 units, respectively, during the same period. Powder bed fusion and binder jetting are the most common processes used by leading vendors, more of whom (70 percent) use metal than any other material.[24] Table 1 presents several leading vendors that manufacture machines, an overview of processes and applications, and the most frequently used materials for each process.

Box 1. Selective Laser Sintering: an Additive Manufacturing Process

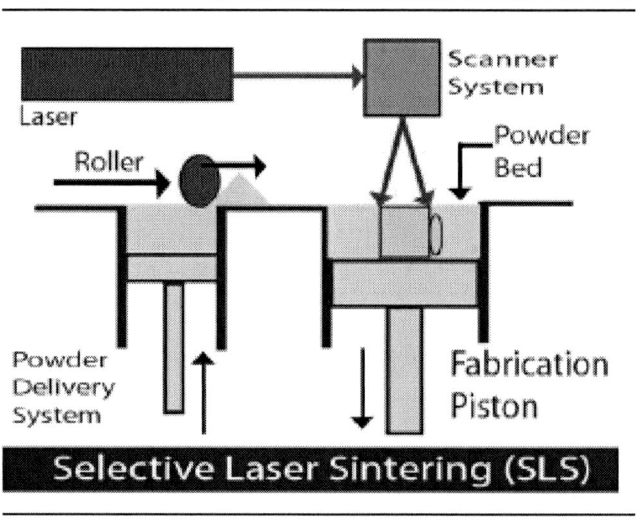

Source: Martello Co., http://www.martello.co.uk/rapid_prototyping.htm (accessed August 3, 2013).

Selective laser sintering uses a high-powered pulse laser to fuse particles of plastic, metal, or glass powders into a desired three-dimensional shape. The laser selectively fuses powdered material by scanning cross sections generated from a 3-D digital description of the design or object. After each cross section is scanned, the powder bed is lowered by one layer of thickness, and a new layer of material is applied. The process is repeated until the object is complete.

Table 1. Leading industrial additive manufacturing machine vendors, 1988-2011

Vendors/Production Sites	Processes/Applications	Materials
3D Systemsa (US, AUS, NED, ITA)	Binder jetting, material jetting, vat photopolymerization, powder bed fusion	Plastic, polymer, metal
Beijing Tiertime (CH)	Material extrusion	Polymer
DWS (ITA)	Vat photopolymerization	Polymer
Envisiontec (GER, US)	Vat photopolymerization, material extrusion	Biomaterial, ceramic, polymer
EOS (GER)	Powder bed fusion	Ceramic, metal, polymer
ExOnea (US, GER, JPN)	Binder jetting	Ceramic, polymer, metal
Objetb (ISR, US, GER, Asia)	Material jetting	Biomaterial, polymer
Solidscape (US)	Material jetting	Plastic
Stratasysa,a,b (US, GER, IND)	Material extrusion	Polymer
Z Corp. (US)	Powder bed fusion	Plastic, metal

Sources: Wohlers, Wohlers Report 2012, 2012, 135, 136; "Introduction to Additive Manufacturing," Ceramic Industry, December 2012.

[a] Stratysys acquired Solidscape in 2011, and merged with Objet in 2012; 3D Systems acquired Z Corp. in 2012.

[b] Also fabricates parts.

Additive manufacturing offers industry a range of unique possibilities. The technology makes it possible, for example, to produce viable three-dimensional objects of virtually any complex geometry without significantly increasing the cost of the parts.[25] Additive manufacturing also has the potential to transform the rules of design by reducing—and perhaps even eliminating—the constraints of molds and dies.[26] Major industrial sectors additively manufacture parts ranging from visual aids to patterns for metal casting. Due to the speed and efficiency with which additive manufacturing can produce prototypes and parts,[27] the technology will have the greatest impact on products that require high customization, have complex designs, and are made in small quantities.[28]

However, additive manufacturing is not yet suitable for mass production. Inherent limitations in the processes include lengthy build time: for example, additive manufacturing processes are capable of creating a 1.5 inch cube per hour, on average, while an injection molding machine can produce several similar parts in under a minute. Current additive manufacturing technologies are unlikely to be able to create parts as quickly as molding technologies.[29] Other limitations include the size of the objects that can be made, and the cost and size of the machines and materials used.[30]

U.S. Additive Manufacturing Market and Major Industrial Sectors

The United States has several advantages in manufacturing in general, and in additive manufacturing in particular. For example, U.S. R&D spending for total manufacturing in 2011 was $415.0 billion, the highest among the countries for which Structural Analysis (STAN) data from the Organisation for Economic Co-operation and Development are available.[31] Additionally, the technology was predominantly developed in the United States, where several leading producers of additive manufacturing machines, including Stratasys and 3D Systems, are based. In 2011, the United States accounted for 38.3 percent of the cumulative installed industrial additive manufacturing systems (Figure 1).[32] The same year, the United States accounted for 64 percent of all industrial additive manufacturing systems sold worldwide.[33]

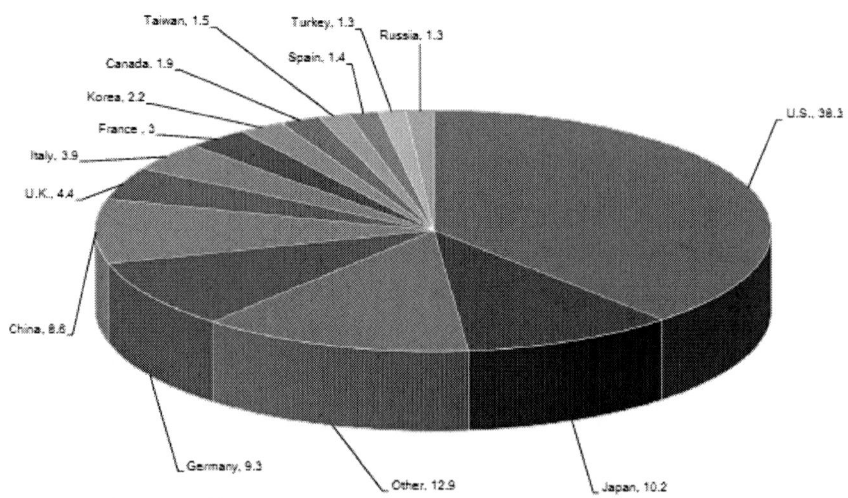

Source: Wohlers, Wohlers Report 2012, 2012, 20.

Figure 1. Cumulative additive manufacturing machines, installed by country, 1988-2011, (percent of global total).

Leading sectors in this field, based on revenue for all additive manufacturing products and services in 2011,[34] include automotive (19.5 percent), medical (15.1 percent), and aerospace (12.1 percent) (Figure 2).[35]

Information to show how broadly additive manufacturing is being deployed in the U.S. manufacturing base, and, in what specific sectors, is not yet available. However, data from the U.S. Department of Commerce show that goods using additive manufacturing methods represent less than a fraction of 1 percent of their relevant industry subsectors, implying abundant potential for growth (table 2).[36]

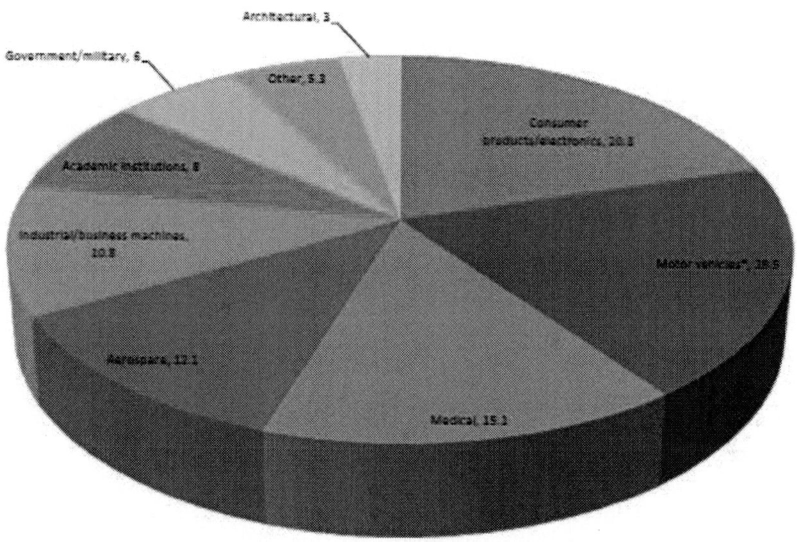

Source: Wohlers, Wohlers Report 2012, 2012, 20.

Figure 2 Industries served and approximate revenues (by percent) for additive manufacturing, 2011.

Automotive

The automotive industry has used additive manufacturing to make tool prototypes and small custom parts with short production runs (in the hundreds) for decades. It is increasingly applying the technology to metals, with a focus on aluminum alloys, to construct lightweight vehicles. Additive manufacturing shipments for the U.S. automotive industry were valued at $48 billion in 2011. As noted earlier, approximately 19.5 percent of additive manufacturing occurs within the automotive industry, with additive manufacturing shipments estimated to be less than 0.05 percent of total U.S. automotive shipments.[37]

Table 2. U.S. Additive manufacturing (AM) shipments, 2011

Category	Relevant industry NAICS codes	Shipments of US-made AM products (billion $)[a]	Total industry shipments (billion $)	AM share of total industry shipments (percent)[b]
Motor vehicles	3361, 3362, 3363	0.048	445.3	0.01
Aerospace	336411, 336412, 336413	0.030	157.7	0.02
Industrial/business machines	333	0.027	365.7	0.01
Medical/dental	3391	0.037	89.5	0.04
Government/military	336414, 336415, 336419, 336992	0.015	32.8	0.05
Architectural	3323	0.074	72.2	0.01
Consumer products/ electronics, academic institutions, and other	All others within NAICS	0.083	895.7	0.01
Total	332 through 339	$.25	$2,058.9	0.01

Source: Thomas, Economics of the U.S. Additive Manufacturing Industry (prepublication draft), June 2013.

[a] These values are calculated assuming that the percent of total additive manufacturing made products for each industry is the same for the U.S. as it is globally. It is also assumes that the U.S. share of AM systems sold is equal to the share of revenue for AM products.

[b] If rounded to "1" right of decimal.

Note: Numbers may not add up to total due to rounding.

*Note: For the purposes of this article, motor vehicles and automotive are used interchangeably.

**Note: The primary additive manufacturing market consists of all products and services directly associated with additive manufacturing processes, worldwide.

An example of an early adopter within the industry is Ford Motor Company, which has been using the technology to develop prototype parts for test vehicles since the 1980s. With industrial-grade machines that cost as much as $1 million each, Ford engineers have produced prototypes of cylinder heads, brake rotors, and rear axles in less time than traditional manufacturing would require. The company reports that use of the technology saves, on average, one month of production time when creating a casting for a prototype of a complex cylinder engine head that includes multiple ports, ducts, passages, and valves to manage fuel and air flow.[38]

Additive manufacturing for automotive parts is also showing increasing potential. For instance, Daimler AG recently funded the development of the X Line 1000R System, a laser fusing machine with a large build volume[39] that uses powdered metals to additively manufacture components for vehicles and engines. The machine, which was introduced at the Euromold Trade Show in November 2012, has a build chamber of 23.6 inches x 15.7 inches x 19.7 inches and a layer thickness of 20 to 100 microns. The new machine has a high-powered laser in the kilowatt range, enabling as much as 10 times more productivity than standard laser fusing machines used in additive manufacturing.[40]

Automotive companies are applying additive manufacturing to an expanding range of parts, including engines and vehicle bodies. An extreme example is KOR EcoLogic's Urbee, 60 percent of which will be additively manufactured. The frame, chassis and engine will be conventionally manufactured from metal.[41] In March 2013, KOR EcoLogic announced that it is collaborating with Stratasys, Ltd., a maker of additive manufacturing equipment, to produce the Urbee 2, a road-ready, fuel-efficient car.[42] The two-passenger vehicle will be able to travel 70 miles per hour on the freeway, and use a biofuel such as ethanol.[43]

When additive manufacturing can produce larger components, the automotive industry will likely adopt the technology more broadly. Like the aerospace industry, as discussed later, the automotive industry has high demands for performance and weight reduction that additive manufacturing can help to address.

Medical

The medical industry is another leading user of additive manufacturing. Additive manufacturing is being used to create customized medical devices that closely replicate the human form. Approximately 15.1 percent of U.S. additive manufacturing takes place within this industry, with additive manufacturing shipments estimated to be $37.2 million—again, less than 0.05 percent— of medical and dental manufacturing.[44] It is the dominant technology for manufacturing customized in-the-ear hearing aids, which have dramatically increased patient comfort. Additive manufacturing has also been used worldwide to create approximately 30,000 prosthetic limbs, more than half a million dental implants, and countless other devices.[45] Shipments for

U.S. manufacturing of medical and dental products amounted to $89.5 billion in 2011.[46]

One of the most well-known applications of additive manufacturing is InvisAlign orthodontics. Align Technology, Inc. uses stereolithography, a form of vat polymerization, to custom-make a series of clear plastic aligners that correct malocclusions[47] and straighten teeth without the use of metal or ceramic braces and wires. Patients wear sequential liners, which they change about every two weeks, until treatment is complete. In 2012, Align Technology shipped 363,540 cases of aligners and, as of March 2013, had trained more than 78,000 dentists and orthodontists worldwide to use the technology to fabricate products on-site.[48]

In recent years, scientists have devised a means of using a patient's cells to additively manufacture skin tissue and an array of other human body parts. In February 2013, doctors at Weill Cornell Medical College and biomedical engineers at Cornell University announced that they had used additive manufacturing and injectable gels made of live cells to build a facsimile of a human ear.[49] According to one industry expert, "the ability to manufacture living human tissue for medical research and clinical practice has the potential to reshape the future of medicine."[50]

The pharmaceutical industry is also applying the technology. For example, additively fabricating a single custom-made daily pill for each patient eliminates the need to keep track of multiple medications. Identification markings on the pill can eliminate the confusion and uncertainty of conventional medicines.[51]

While there are no formal projections of future use, experts anticipate new machines and new processes will expand applications, particularly in orthopedic implants.[52]

Aerospace

Additive manufacturing holds significant potential for the aerospace industry, which requires parts that are lightweight, strong, and geometrically complex—and typically produced in small quantities. Total (both traditional and additive) manufacturing shipments in the U.S. aerospace industry were estimated at $157.7 billion in 2011. Aerospace accounts for approximately 12.1 percent of additive manufacturing, with the total shipments estimated to be $29.8 million, or less than 0.05 percent of the U.S. aerospace industry.[53]

Aerospace firms are increasingly turning to the technology to reduce the costs of developing models and prototypes and of creating components. In a constant effort to reduce aircraft weight,[54] the industry is developing a growing proportion of its parts from titanium, plastic, and other lightweight materials. Many of these materials are costly, and additive manufacturing can make it possible to keep the amount used to a minimum. Aircraft landing gear, for example, can be additively manufactured from titanium layer by layer—rather than cut from a titanium block—thereby greatly reducing material waste and costs.[55] In addition, additive manufacturing uses these materials to create parts in geometries that may not be possible with conventional manufacturing. It can also significantly reduce the waste generated in producing the materials themselves, as well as in manufacturing and using the components.[56]

The increasing breadth and sophistication of these applications are, in turn, driving needs for improvements in process control, materials, and inspection to ensure safety.[57] Boeing, Inc., for example, has additively manufactured more than 200 different parts for 10 aircraft platforms.[58] Boeing has also used roughly 20,000 additively manufactured parts in military and commercial aircraft, including 32 different components for its 787 Dreamliner planes.[59] General Electric, the world's largest supplier of jet engines, is preparing to additively manufacture a fuel nozzle for use in thousands of jet engines.[60] These nozzles are 25 percent lighter and as much as five times more durable than the existing model, which is welded from 20 different parts.[61] General Electric reports that it may additively manufacture up to half of the parts in its energy turbines and aircraft engines in 10 years.[62]

Aurora Flight Sciences, in partnership with Stratasys, Ltd., fabricated and flew a 62-inch wingspan aircraft with a wing composed entirely of additive manufactured components. The wing was designed by Aurora and manufactured by Stratasys utilizing fused deposition modeling 3D printers.[63] Aurora anticipates additively manufacturing small, unmanned aerial vehicles— both military and civilian—in five years.[64] In a public-private venture, the National Air and Space Administration (NASA) and Pratt & Whitney Rocketdyne (PWR) jointly produced an additively manufactured rocket engine injector. A rocket engine injector is one of the most critical and expensive engine components in a launch vehicle, and typically takes up to a year to conventionally manufacture. The NASA-PWR team was able to additively manufacture it in four months while reducing costs by 70 percent.[65]

Despite these advances, limits on the size of goods produced by additive manufacturing have likely constrained adoption of this technology in the aerospace industry. Issues with materials, accuracy, surface finish, and certification standards have further limited its use.[66]

Beyond these aerospace/medical/automotive uses, additive manufacturing has applications in many other industries, ranging in size from miniature instrumentation to large structures such as the earth boring drills used in oil exploration. The applications are so broad that they cannot be easily classified, but several are seen on a recurring basis, such as specialized manufacturing fixtures and tools, power generation equipment, robotics, heat exchangers, and thermal controls.[67]

The Potential Benefits of Additive Manufacturing

Additive manufacturing competitively produces low-volume, customized, and intricate goods because it enables inexpensive design of prototypes and parts. Additional levels of product complexity do not add costs to production beyond the design stage, because once the design is set, complex layering is no more expensive than if you were to layer a cube or a sphere.

In traditional manufacturing, by contrast, the prototype, tooling, and setup processes can be too time-consuming and expensive to profitably produce small volumes. In addition to these advances, additive manufacturing technology is fostering innovation and transforming how some companies produce and deliver goods and services. Firms are beginning to employ additive manufacturing as a tool to achieve objectives such as increasing supply chain efficiencies; reducing time to market; moving from mass production to mass customization; and sustaining the environment.

Supply Chain Efficiencies

Additive manufacturing has the potential to reduce the costs of storing, moving, and distributing raw materials, mid-process parts, and end-usable parts. The ability to produce parts on demand without the need for tooling and setup could become a basis for new solutions in supply chain management. Ed Morris, director of America Makes, the federally-funded initiative set up to define and promote the future of the industry, reports that “you don’t have to have the finished product stacked on shelves or stacked in warehouses anymore... . Whenever you need a product, you just make it. And that collapses the supply chain down to its simplest parts, adding new efficiencies to the system.”[68] As noted, General Electric is now additively manufacturing hollow fuel nozzles for aircraft that it previously manufactured in pieces and assembled later. The company has thus reduced the product's supply chain.

Reducing Time to Market

Time to market is expected to shrink in additive manufacturing applications due to shorter design and prototyping cycles, more predictable factory loads, and the elimination of tooling and factory setup times for new products. The freedom to design and redesign prototypes and parts without slowing down or adding to production costs enables a more fluid product development process. Similarly, the ability of machines to read CAD files improves planning: machines know the time and material requirements necessary to build a part before it is on the machine, and can track and measure volume and capacity at any moment.[69] For example, General Motors used additive manufacturing to significantly redesign models for the 2014 Chevrolet Malibu at a much lower cost than for its previous clay sculpting processes.[70] The company reports the technology was particularly useful for updates to the new Malibu's floor console, which features a pair of integrated smartphone holders for driver and passenger. The redesigned console also weighs less, which contributes to the car's improved fuel efficiency.[71]

Mass Customization

Additive manufacturing's ability to employ multiple designs on the same machine could enable the manufacturing industry to move from mass production in factories to mass customization with distributed manufacturing.[72] Using materials ranging from plastic to titanium to human cells, additive manufacturing creates intricate products of a near-infinite variety that can be made to exact customer specifications.

Bespoke Products, a division of 3D Systems, additively manufactures custom-designed prosthetic human body parts, such as legs. The company builds models of complete legs that have sophisticated features such as body symmetry, locking knees, and flexing ankles. During the development process, customers select from a range of options for customizing their product. The additively manufactured legs cost $5,000 to $6,000 each, and have features that are not available in existing prosthetic legs, which can cost up to $60,000.[73]

Environmental Sustainability

Additive manufacturing could become a multifaceted tool for mitigating environmental impact by replacing many of the casting, molding, and other manufacturing processes that consume significant amounts of energy and produce hazardous industrial waste. The technology also imposes few constraints on product design, enabling previously separate parts to be consolidated into a single object with increased functionality while reducing the amount of energy and natural resources required to operate it. The U.S. Department of Energy reports that energy savings of 50 percent or more can be realized in applications where additive manufacturing is competitive.[74]

PRESENT TRENDS

Numerous, multifaceted dynamics are shaping additive manufacturing. As users demand more from the technology, developers are investing in increasingly advanced processes and materials. Indeed, many industry experts see additive manufacturing as a way of providing high-value, high-margin parts and products that has a potential for explosive growth. Others, however, point to weaknesses in the technology that they argue will continue to challenge development and adoption of additive manufacturing for the foreseeable future.

Drivers to Development and Adoption

A mix of market forces and technologies are driving development and adoption of additive manufacturing. Three of the most significant ones are mass customization, new and improved processes and products, and government initiatives.

Mass Customization

Mass customization is a significant driver—as well as a result—of additive manufacturing.[75]In response to increasingly fragmented markets that value individualization, manufacturers across industries are using innovative manufacturing techniques and technology to both mass-produce and individually customize products.[76] In the automotive industry, additive manufacturing enables companies to quickly and distinctively design cars, and rapidly customize them. Ford Motor Company allows its customers to "build" certain aspects of a vehicle by choosing from a palette of online additively manufactured options, such as body side cladding.[77] BMW North America also provides the option to customize several of its vehicles.[78] Customers can choose from a wide variety of additively manufactured features, including trims and inlays.

The medical device industry is similarly using the technology to build customized implants and devices. The Walter Reed National Military Medical Center uses the technology to produce customized cranial plates and cutting guides for bone grafts that are less expensive than existing alternatives and better matched to the patient.[79] Additionally, 95 percent of the world's custom "in the ear" hearing aids are additively manufactured.[80]

Additive manufacturing offers numerous advantages for mass customization. Flexibility, speed, and the ability to build objects directly make it an attractive alternative to conventional machining, forging, molding, and casting of customized parts. However, business models are only now catching up with the movement toward greater mass customization. Companies are still determining how to capture individual or group customer requirements, how much input customers will have in the design process, and which parts of the product will add the most value to the customer when customized. Given the trend, the reciprocal relationship between additive manufacturing and mass customization will likely accelerate.[81]

New and Improved Processes and Products

As noted, advances are occurring in basic areas, such as manufacturing machines, and in more applied areas, such as nanotechnology.[82] Below are several of the most significant developments in additive manufacturing processes and the end products they create, as identified by the Institute for Defense Analysis.

Incorporating Energy and Electronics

Advances are being made in additively manufacturing conformal electronics.[83] Robotic dispensing systems could apply conformal coating to printed circuit boards to protect them against moisture, dust, chemicals, and extreme temperatures. The systems would be able to create a repeatable coating thickness and a precise coverage area that provides superior quality and reduces costs. While no commercially available additive manufacturing process currently produces conformal electronics, the concept has been demonstrated in a joint project between Sandia National Laboratory and the

University of Texas at El Paso.[84] In addition, researchers at Cornell University have additively manufactured parts that embed electronics, such as conductors and zinc batteries.[85]

Creating New Structures

Additive manufacturing enables the creation of structures with distinct advantages over traditionally manufactured ones: they are more complex, have greater geometric freedom, and are more capable of performing multiple functions. The technology allows for characteristics such as nonlinear holes and re-entrant features on complex parts with little—if any—addition to cost.[86]It also enables structures, such as honeycombs and lattices for the interiors of parts, which have significant benefits: maintaining the requisite strength and stiffness while being considerably lighter than conventionally manufactured counterparts. Researchers at the Lawrence Livermore National Laboratory of the U.S. Department of Energy are additively manufacturing combinations of materials in new ways and creating materials with properties not found in nature that will further innovation and design capabilities.[87]

Three-Dimensional Scanning

New three-dimensional scanners and processing software options are expanding opportunities for designing and additively manufacturing prototypes and parts. New algorithms, for example, are able to transform relatively crude scans into quality three-dimensional surface mapping.[88] These developments hold great potential for fields such as robotics and human-computer interaction.[89]

Bioprinting

Advances in custom medical implants and devices are driving demand for additive manufacturing. As discussed, the Walter Reed National Military Medical Center has successfully produced and implanted porous cranial plates and cutting guides for bone grafts that are less expensive than existing alternatives and better matched with the patient.[90] Progress has also been made in additively manufacturing soft tissue and organs such as kidneys.[91] Heart valves and small veins have also been made, but are not yet ready for implementation.[92]

Government Initiatives and Public-Private Partnerships

As part of an effort to strengthen productivity and competitiveness in manufacturing, national governments around the world have launched several initiatives to promote additive manufacturing. China, Singapore, and Germany are investing heavily in the technology (box 2). The White House Office of Science and Technology has formed an interagency working group, and numerous federal organizations, such as the Department of Commerce, the National Science Foundation, and the Department of Energy, have programs to advance the technology. For example, engineers at the National Institute of Standards and Technology (NIST) are working to enhance additive manufacturing equipment and process metrology; process optimization and control; advanced sensor systems; materials characterization; data formats; and standards development.[93]

The U.S. federal government is also working with private industry and academic institutions to further advance additive manufacturing. NIST awards grants to improve additive manufacturing; it provided $7.4 million of funding in 2013.[94] The National Science Foundation has programs to educate students, workers, and enterprises about additive manufacturing. It also awards grants to improve the technology. The Department of Energy Advanced Manufacturing Office partners with industry, small businesses, and other stakeholders to identify and invest in additive manufacturing technologies. With the Oak Ridge National Laboratory, it has a program to develop advanced materials; implement advanced controls; and explore next-generation systems to overcome technology barriers for additive manufacturing. One project involves the development of large-scale, high-deposition-rate additive machining centers.[95]

Box 2. Initiatives in Other Countries

Industrialized economies, and a few developing ones, are working to develop and apply additive manufacturing, and thus strengthen productivity and competitiveness.

China, Singapore, and some countries in Europe have committed hundreds of millions of dollars to develop and commercialize additive manufacturing. China, for example, has been investing in additive manufacturing since the early 1990s, and the Chinese government is pledging 1.5 billion yuan ($245 million) to a seven-year project to advance development of the technology.a The Asian Manufacturing Association,[a] Beijing-funded trade group, is promoting wider integration of additive manufacturing by establishing 10 innovation institutes, each starting with a $3.3 million injection of investment.[b] A company in Hefei, Anhui Province, is investing 750 million yuan ($125 million), indicating that Chinese business is eager to explore the technology.[c]

Singapore is investing $500 million in additive manufacturing over the next five years.[d] The Singaporean government is dedicating significant funding to specific programs to help engineers define and use next-generation manufacturing technologies, and to explore the potential of building a new additive manufacturing industry ecosystem.[e]

There are also numerous efforts to further additive manufacturing in Europe. One initiative is the Direct Manufacturing Research Center at the University of Paderborn in Germany. Established in 2008, this institution is a leading research center for additive manufacturing, with funds from the German state of North Rhine-Westphalia and member companies. Research has thus far focused on improving the processes for laser sintering/melting technology for metal and plastic powder, as well as establishing industry requirements for materials, training, and standards development.[f] In October 2012, the UK government announced grants for collaborative research and development projects in additive manufacturing. The grants will be awarded through an open competition managed by the Technology Strategy Board, the Engineering and Physical Sciences Research Council, and the Economic and Social Research Council.[g]

[a] Brooke, "China Flexes Muscles," June 27, 2013.
[b] Brooke, "China Flexes Muscles," June 27, 2013.
[c] With the exception of Japan, additive manufacturing started much later in Asia than in the United States and Europe. Asia has largely caught up with the rest of the world; however, most industrial machines there are bought by product development companies to assist in early product development rather than final goods. Brooke, "China Flexes Muscles," June 27, 2013.
[d] Brooke, "Singapore to Invest," March 27, 2013.
[e] Brooke, "Singapore to Invest," March 27, 2013.
[f] Direct Manufacturing Research Center website, http://dmrc.uni-paderborn.de/about-dmrc (accessed July 5, 2013).
[g] The Engineer, "£7m Funding for UK Additive Manufacturing Projects," October 23, 2012.

America Makes, formerly the National Additive Manufacturing Innovation Institute (NAMII)— a public-private partnership with member organizations from government, industry, and academia—is a particularly robust additive manufacturing program.[96] America Makes is the flagship of 15 manufacturing institutes developed by the National Network for Manufacturing Innovation, a billion-dollar initiative to strengthen U.S.

manufacturing and competitiveness.[97] The stated mission of America Makes is "to accelerate additive manufacturing technologies to the U.S. manufacturing sector and increase domestic manufacturing competitiveness."[98] It is promoting the technology by:

- establishing a collaborative network for sharing information and research;
- facilitating the development and application of efficient additive manufacturing technologies; and
- developing a workforce that can adapt to the changing requirements of additive manufacturing technologies and practices.[99]

America Makes is the first national program dedicated to additive manufacturing. It was launched in 2013 with $90 million in funding: $30 million from the federal government and the rest from the business and nonprofit sectors. The 20 projects underway range from basic research about how polymers and other materials will react during the heating and deposition process to work on more industrial applications, such as developing a lower-cost, high-temperature process for working with thermoplastics to make air and space vehicle components.[100]

Industry is taking steps to facilitate the development of additive manufacturing.[101] EWI, a nonprofit engineering and technology resource organization, launched the Additive Manufacturing Consortium (AMC) in 2010. The AMC is a diverse group of practitioners and stakeholders working to "accelerate the innovation in additive manufacturing technologies to move them into the mainstream of manufacturing technology from their present emerging position."[102] Stratasys, Ltd. is working with the Oak Ridge National Laboratory to develop fused deposition modeling[103] and make it a mainstream manufacturing practice.[104] Firms are also partnering with academia to promote additive manufacturing. Jet engine manufacturer Aerojet Rocketdyne, for example, is collaborating with the University of Connecticut to create one of the country's most state-of-the-art additive manufacturing facilities. Over the next five years, Aerojet Rocketdyne will invest $9 million in the development of powder-based additive manufacturing technologies.[105] These and other efforts aim to accelerate U.S. development and application of additive manufacturing technology and will likely help strengthen U.S. competitiveness.

Challenges to Development and Adoption

Several fundamental challenges must be addressed to achieve widespread use of additive manufacturing and to realize its potential economic benefits. Among the most significant challenges are developing standards, improving the selection and affordability of materials, and increasing the reliability and accuracy of equipment and processes.[106]

Standards Development

Developing standards is critical to increasing diffusion and adoption of the technology.[107] Standards would provide a foundation for creating products that conform to certain specifications and are compatible with products provided by different suppliers seeking the same quality, performance, and interchangeability.[108] Additive manufacturing standards would also ensure the safety, reliability, and quality of processes and products. Because there are currently only a handful of additive manufacturing standards, companies conduct their own testing to ensure integrity of the equipment, processes, and products.[109] Costly and time-consuming, testing deters wider application of additive manufacturing, underscoring the need to develop standards from design to part build to operation. A standard is needed, for example, for software compatibility and communication across additive manufacturing machines. Instead of a standard format for electronic blueprints that allows CAD programs to talk to each other, there are a number of proprietary and ad hoc formats that do not work well together, slowing and limiting production.[110]

Another area of additive manufacturing production that would benefit from standards development is material property data generation.[111] Measurement (metrology) capacity, among others, is severely lacking. Experts report that there is a "striking dearth of metrology capabilities for process-structure-property relationships; closed-loop and adaptive control systems; and new sensors for fundamental build properties such as shape and surface finish."[112] They further note critical needs in measuring the raw material inputs, processes, and resulting parts.[113]

All material property data must be robust and comprehensive. In 2009, in response to the need for industry standards, the Society of Manufacturing Engineers and ASTM International (formerly the American Society for Testing Materials) formed Committee F42 on Additive Manufacturing

Technologies. The Committee, of which NIST is an executive member, is working to develop standards to:

- allow manufacturers to measure and compare the performance of different additive manufacturing processes and materials;
- specify part-building requirements to give purchasers and suppliers a common set of parameters with which to work, thereby improving vendor relationships;
- help new users adopt additive manufacturing technologies; and
- provide users with uniform procedures for calibrating additive manufacturing machines and testing their performance.[114]

The NIST Engineering Laboratory reports that the additive manufacturing industry has identified many additional needs for standards that will take much effort and time to develop.[115]

Material Selection and Cost

The narrow selection of inputs and uncertain properties of additively manufactured materials present barriers to many industries. This is especially true in aerospace, which needs material that can withstand staggering pressures and temperatures of over 2,000 degrees Fahrenheit.[116]

Despite recent progress in additive manufacturing material development and characterization,[117] problems remain, including the limited selection of materials, inconsistent production quality, and high costs. As a result, only a fraction of the materials that are used in conventional manufacturing are compatible with additive manufacturing.[118] As noted earlier, the major categories of additive manufacturing materials are plastics and other polymers and metals. There are also diverse filled and composite materials, as well as ceramic and ceramic-metal hybrids.[119] According to NIST, "material requirements are impacted by the need to create feedstock, to be processed successfully by the fabricator coupled with post processing, and to manifest acceptable service qualities."[120]

Most additive manufacturing processes use proprietary polymers that are not well characterized, are often weaker than those used in traditional manufacturing, and can sometimes lack uniform part strength. Parts that have been additively manufactured with metal have physical properties that can be quite different from conventional wrought or cast metals. They may, for example, lack full density, which can compromise fracture toughness and fatigue properties. According to an industry expert, "porosity, or partial delamination, could act as crack initiation sites in the material, which could lead to premature failure of such parts, especially when subjected to cyclic stress conditions."[121]

Although there have been increases in the variety and application of additive manufacturing material inputs and feedstocks, they are still expensive relative to traditional manufacturing materials. For example, powder metals can be 200 times as costly as sheet metal,[122] and photopolymers cost $750–$1,000 per gallon, compared to injection molding material, which costs $1 per pound.[123] While some producers enjoy savings when using additive manufacturing for custom products and low production runs, high material prices continue to be an impediment for many potential producers.

Equipment and Processes

Establishing additive manufacturing as a mainstream technology is, in part, contingent upon overcoming significant constraints in equipment and processes. Limitations such as machine cost and software capability will likely be overcome in the medium term.[124] Indeed, prices of industrial machines are falling, and capabilities of software and scanning devices are improving. Other limitations, however, are less likely to be surmounted soon.[125] The physics of layering and curing, for example, impose limits on process speed. Additionally, the issue of minimum wall thickness[126] is a fundamental restraint that is unlikely to be resolved without significant technological advancement.[127]

The *Roadmap for Additive Manufacturing*,[128] developed by experts in academia, government, and industry, outlines the following technological barriers:

- High costs of machinery and materials in comparison to those used in conventional manufacturing;
- Product fabrication speeds that are significantly slower than mass production processes;[129]
- Inherent tradeoffs between product size, accuracy, and speed (e.g., larger surfaces entail lengthier production times; accelerated production processes may produce parts with unknown and unpredictable properties);

- Significant geometric and property variations between "identical" products built on different machines leading to a lack of repeatability (e.g., use of the same materials and identical CAD files on five different machines could produce five goods with different properties);
- Requirements for highly skilled operators and/or careful periodic tuning, both of which are in short supply and costly;
- Unreliable machinery and inconsistent product quality, particularly with regard to part accuracy and surface finishes;
- Closed architectures, which preclude researchers from making meaningful changes to processing conditions; and
- The lack of hardware and software necessary for simple and effective multi-material deposition.[130]

INDICATORS OF FUTURE GROWTH

Increasing sales, technological advances, new applications, and the expiration of key patents are indicators of likely continuing growth in additive manufacturing. They are also drivers of the market, fueling momentum in investment, development, and applications.

The world market for additive manufacturing equipment and services has steadily and rapidly expanded since the mid-1980s. Wohlers Associates reports that in 15 of the 24 years it has tracked the technology, revenue[131] in the primary additive manufacturing market grew in the double digits and, in 2011, increased an estimated 29.4 percent to $1.7 billion[132] (Figure 3). This estimate comprises all products and services directly associated with additive manufacturing processes worldwide.[133] Product revenue of $834 million includes additive manufacturing machines, machine upgrades, materials, and aftermarket products such as third-party software and lasers.[134] Services revenue includes $642.6 million from parts produced on additive manufacturing machines by service providers, such as replacement components and $236.9 million from machine maintenance contracts and activities such as training, seminars, conferences, contract research, and consulting.[135]

Revenue is expected to continue strong double-digit growth over the next several years, with sales of additive manufacturing products and services likely reaching $3.7 billion worldwide by 2015 and surpassing $6.5 billion by 2019.[136]

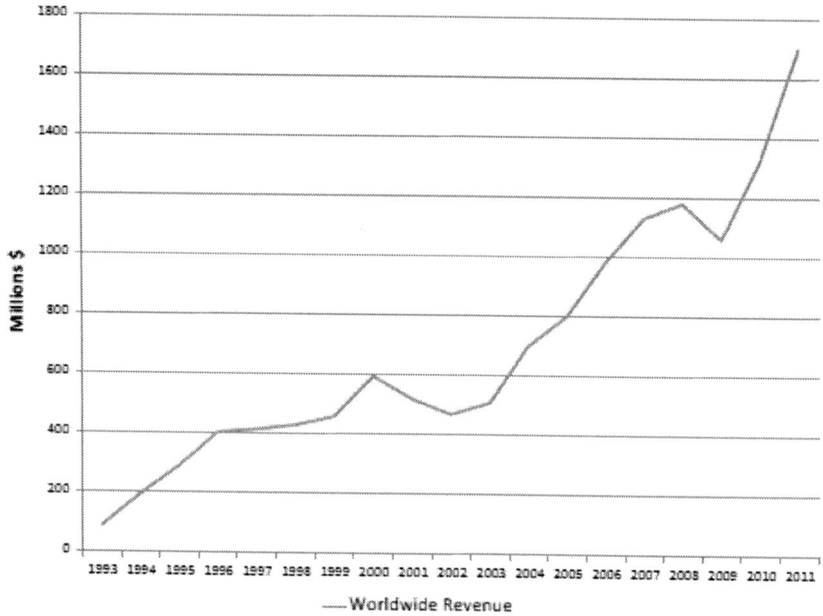

Source: Wohlers, Wohlers Report 2012, 2012, 125-127.

Figure 3. Worldwide revenue for additive manufacturing products and services, 1993–2011.

The United States leads in the production and sales of professional-grade industrial additive manufacturing machines, accounting for 38 percent of installed machines during 1998–2011 (Figure 2). Nearly two-thirds of all the industrial machines sold in 2011 came from U.S. manufacturers.[137] Wohlers Associates reports that annual sales of professional-grade, industrial additive-manufacturing machines will exceed 10,000 units worldwide by 2016.[138] Machines priced in the $5,000–$25,000 range will be responsible for most of this growth.[139] The growth reflects industry developments in and increasing awareness of additive manufacturing.

Advances in Materials

As discussed above, the range of feedstock applications available for use in additive manufacturing pales in comparison to that for traditional manufacturing. Over the past five years, however, significant improvements have occurred in the properties, variety, and quality of the applications for additive manufacturing.[140] It is now possible, for example, to additively manufacture parts in a range of metals that are on par with wrought material and exceed the properties of castings.[141] Among the available metals for additive manufacturing are aluminum alloys, titanium alloys, nickel-based superalloys, and a range of steels. Engineers are able to additively fabricate fully functional components from titanium and various steel alloys, featuring material properties that are equivalent to their traditionally manufactured counterparts. "Direct metal" processes will continue to advance as process control and understanding of fundamental metallurgy for additive manufacturing improves. Recent work at the Fraunhofer Institute for Laser Technology points to the potential for a fourfold increase in printing speeds for metal objects.[142]

The increase, if achieved, would make additive manufacturing more attractive to a wider range of industries—and more competitive with traditional manufacturing.

The variety and quality of plastic, ceramic, and hybrid materials for additive manufacturing also continues to increase. EOS GmbH recently released three new polymers for its powder bed fusion process. PrimePart and the French chemical company Arkema have begun marketing two nylon-based powders designed for easy processing.[143] Scientists at General Electric Global Research and designers at General Electric Aviation have developed a group of new materials, ceramic matrix composites (CMC). Additively manufactured CMC components outperform the most advanced metallic alloys in gas turbines for jet aircraft.[144] CFM International, a joint venture between General Electric Aviation and France's Snecma, completed designs for a jet engine that includes additively manufactured ceramic matrix composites. CFM plans to build the first engine in 2014, and to test and certify it over the next two years.[145]

As a result of breakthroughs in multi-material additive manufacturing, more complex materials are being produced. The Objet Connex500, for example, additively manufactures up to 14 plastic-like materials in one job run. Rather than being fabricated as separate components and attached one at a time, they are simultaneously fused together. Researchers are developing new materials such as high-temperature polymers integrated with optics gradient materials, in situ sensors,[146] self-assembled single crystals, and continuous-filament composites.[147] The growing number and variety of materials for additive manufacturing increases the range of possible users and applications.

New Design Tools

New design tools specific to additive manufacturing are proliferating. Until recently, inputs into an additive manufacturing machine were limited to a CAD model. Now, input can be generated by a number of means, including medical scan data, entertainment software (as is the case with computer game avatars), and even simple drawing and sketching programs.[148]

Software and design tools such as Autodesk 123D Suite are becoming increasingly popular.[149] The growing sophistication of the tools will enable production of new, complex designs as well as the analytical techniques to validate the designs before manufacturing.

Future design tools will be created to be intuitive to use. Some, for example, will respond to touch and movement; others will respond to environmental conditions.[150] Co-design and co-creator tools are also emerging and can often be accessed and used on a website. With the advent of more advanced design tools, the benefits of additive manufacturing, such as prototyping, can be more fully realized and may spur new products and innovations.[151]

Expiration of Key Patents

The expiration of early additive manufacturing machinery patents is influencing the development of new machines and applications in the United States and abroad. When the original fused deposition modeling patent expired in 2005, it allowed for the creation of RepRap, a replicating rapid prototyper. RepRap was the first low-cost additive manufacturing machine and it has found great popularity in the open-source community. Today, RepRap is more widely used than any other additive manufacturing system.[152]

The last of the selective laser sintering[153] patents from inventor Dr. Carl Deckard and the University of Texas at Austin expired in June 2014. Wohlers Associates report that plans to produce a more affordable alternative are already underway. They suggest that there will be several new developments based on laser sintering technology by mid-2014.[154] As more patents expire, there will be opportunities to capitalize on the technology and to develop new systems. These new systems will likely provide additional product offerings and increase competition among manufacturers.

Potential Implications for Manufacturing Competitiveness

Additive manufacturing is increasingly used in the automotive, medical, and aerospace industries, but largely remains limited to small-series production within them. While the technology has room to grow and potential to strengthen productivity in many industries, its implications for overall manufacturing competitiveness are less clear. As with any emerging technology, new standards, wider awareness, and a better-developed infrastructure are required to facilitate its use.[155] Further, evolving industry and government policy—as well as the convergence of additive manufacturing with other advancing technologies—will certainly bolster the use of additive manufacturing.

Impacts on Manufacturing

It is difficult to predict in detail how additive manufacturing—a technology that is still emerging—will affect individual industries. However, if challenges and barriers noted earlier can be overcome, additive manufacturing has the potential to open the door to myriad new processes and efficiencies. The impacts will initially be limited to a range of customized, small-batch production goods, but there are indications that a paradigm shift has already started. Some likely impacts include:

- reduced material waste in comparison to subtractive methods;
- superior products with complex internal structures that add strength, reduce weight, increase functionality, and are easier to maintain;
- open-design products created by communities of end users; and
- customization as per-unit costs of small production runs (even single items) approach those of long runs.[156]

As additive manufacturing continues to evolve, it will likely result in:

- quicker time to market due to faster design and prototyping cycles as well as the possible elimination of many traditional manufacturing steps such as transportation, tooling, and assembly;

- greater competition, creating a larger variety of products due to lowered barriers to entry; and
- smaller, less costly, and more agile supply chains, especially for low-volume or highly specialized components.[157]

CONCLUSION

Despite the challenges to further adoption, many experts believe additive manufacturing will significantly change certain production and distribution activities. The Economist predicts the technology will have an impact on manufacturing as profound as modern assembly-line factories did in the 20th century.[158] The McKinsey Global Institute notes that "the technology could usher in a new ecosystem of smaller value chains and new companies providing printable designs on the web, instead of products on the shelf... small business will be able to compete with traditional manufacturing, product will never go out of stock, size of batches will become meaningless, and manufacturing will become truly just in time."[159]

The United States is already a leader in the production and use of additive manufacturing. With one of the most innovative and flexible economies in the world, it is also well positioned to further exploit the technology.[160] Additive manufacturing could reduce costs and improve efficiency in manufacturing, which continues to be an important driver of prosperity in the country: manufacturing accounts for 12 percent of U.S. gross domestic product, 70 percent of private R&D spending, 60 percent of exports, and about a third of productivity growth. The establishment of standards and appropriate government and industry policies and initiatives would accelerate development and adoption of the technology. Although the benefits would vary according to industry, the resulting spillover effects would likely enhance the overall productivity and manufacturing competitiveness of the United States.

BIBLIOGRAPHY

Baily, Manyika, and Gupta. "U.S. Productivity Growth: An Optimistic Perspective," Spring 2013, 8. *International Productivity Monitor*. Spring 2013. 8. *http://www.csls.ca/ipm/25/IPM25-Baily-Manyika-Gupta.pdf.*

BioPortfolio. "Additive Manufacturing Technologies for Producing Tissue," July 22, 2013. *http://www.bioportfolio.com/resources/pmarticle/ 383711/ Additive-ManufacturingTechniquesFor-The-Production-Of-Tissue-Engineering-Constructs.htmlonline.*

BoneZone. "The Future of Additive Manufacturing in Orthopaedic Implants," March 5, 2013. *http://www.bonezonepub.com/component/content/ article/689-48-bonezone-march-2013-research-a-development-the-future-of-additive-manufacturing-in-orthopaedic-implants.*

Boeing.com. "Boeing Launches New Manufacturing Venture." News release, September 3, 2002. *http://boeing.mediaroom.com/2002-09-03-Boeing-Launches-NewManufacturingVenture.*

Boulton, Clint. "Barbie, Auto Parts Hot off the Press." *Wall Street Journal*, June 6, 2013.

Bourell, David L, Ming C. Leu, and David W. Rosen, ed. *Roadmapfor Additive Manufacturing: Identifying the Future ofFreeform Processing.* University of Texas at Austin. Laboratory for Freeform Fabrication, 2009. *http://wohlersassociates.com/roadmap2009.html.*

Brooke, Rose. "China Flexes Muscles in 3D Printing Race to the Top." *TCTMagazine*, June 27, 2013. *http://www.tctmagazine.com/additive-manufacturing/china-flexes-muscles-in-3dp-race/.*

———. "Chevrolet Uses Rapid Prototyping to Complete 2014 Malibu," *TCT Magazine*, June 11, 2013. *http://www.tctmagazine.com/additive-manufacturing/chevrolet-uses-rapidprototypingto-complete-2014-malibu/.*

———. "Singapore to Invest $500 Million in 3D Printing." *TCT Magazine*, March 27, 2013. *http://www.tctmagazine.com/additive-manufacturing/ singapore-to-invest500m-in-3dprinting/.*

Brooks, Robert. "Auto Giant Harnesses New Super-scale Laser," *American Machinist*, November 15, 2012, 7. *http://www.industryweek. com/emer ging-technologies/daimler-divesadditive-manufacturing.*

Campbell, Ian, David Bourell, and Ian Gibson. "Additive Manufacturing: Rapid Prototyping Comes of Age." *Rapid Prototyping Journal 18*, no. 4 (2012): 255–58.

Campbell, Thomas A., and Olga S. Ivanova. "3D Printing of Multifunctional Nanocomposites." *NanoToday 8*, no. 2(2013): 119–20. *http://www. sciencedirect.com/science/ journal/17480132/8/2.*

Campbell, Thomas A., and John Slotwinski. "Metrology for Additive Manufacturing: Opportunities in a Rapidly Emerging Technology." *Advances in Engineering Research 7* (2013).

Campbell, Thomas A., Christopher Williams, Olga Ivanova, and Banning Garrett. *Could 3D Printing Change the World? Technologies, Potential, and Implications ofAdditive Manufacturing*. Atlantic Council. Strategic Foresight Report, October2011. *http://www. atlanticcouncil.org/ images/files/publication pdfs/403/101711 ACUS 3DPrinting.PDF.*

Castle Island. "Additive Manufacturing: What RP Will Be When it Grows Up." *http://www. additive3D.com/rm_22.htm* (accessed May 1, 2014).

CSC. "3D Printing and the Future of Manufacturing," Fall 2012. *http://assets1.csc.com/ innovation/downloads/LEF_20123DPrinting.pdf.*

Davidson, Adam. "Making It in America." *Atlantic*, January/February 2012. *http://www.theatlantic.com/magazine/archive/2012/01/making-it-inameri ca/8844/?single page=true.*

———. "The Transformation of American Factory Jobs, in One Company." NPR, January 13, 2012. *http://www.npr.org/blogs/money/ 2012/01/13/ 145039131/the-transformation-ofamerican-factory-jobs-in-one-company?ft=1&f=100.*

Davidson, Paul. "3-D Printing Could Remake U.S. Manufacturing." *USA Today*, July 10, 2013. *http://usatoday30.usatoday.com/money/industries/ manufacturing/story/2012-07- 10/digitalmanufacturing/56135298/1.*

Dehoff, Ryan. "Thinking beyond Today: Comprehending the Potential of Additive Manufacturing Processes, Materials and Properties." Presentation at the National Science Foundation Workshop on Frontiers of Additive Manufacturing Research and Education, Arlington, VA, July 11, 2013. *http://nsfam.mae.ufl.edu/Slides/Dehoff.pdf(accessed* May 5, 2014).

Deloitte. "TMT Predictions 2012." *http://www2.deloitte.com/content/dam/ Deloitte/global/ Documents/Technology-Media-Telecommunications/dttl_ 16264A_TMT_Predict_sg6.pdf*(accessed June 23, 2013).

Destefani, "Ceramic Matrix Composites Make Inroads in Aerospace," May 14, 2013. *http:// ceramics.org/ceramic-tech-today/ceramic-matrix-composites -make-inroads-in-aerospace.*

Economist, "3D Printing," February 14, 2011.

———. "Print Me a Stradivarius," February 10, 2011. —

———. "The Third Industrial Revolution," April 21, 2012.

Engineer. "£7m Funding for UK Additive Manufacturing Projects," October23, 2012. *http://www.theengineer.co.uk/process/news/7m-funding-for-uk-additive-manufacturingprojects/1014350.article.*

Engineering.com. "3D Printed Rocket Blasts Off," July 15, 2013. *http://www.engineering.com/3DPrinting/3DPrintingArticles/ArticleID/60 01/3D-Printed-Rocket.*

Farhadi, Kurosh. "3D Printing, Additive Manufacturing and Drivers for Adoption." Federation of European Screen Printers Associations, October 23, 2012. *http://www.fespa.com/news/ industry-news/3d-printing-additive-manufacturing-and-drivers-for-adoption-fespa.html.*

Freedman, David. "Layer byLayer,"*MIT TechnologyReview*, December 19, 2011. *http://www. technologyreview.com/featuredstory/426391/layer-by-layer/.*

Glass, Nick. "Pitch Perfect: The Quest to Create the World's Smallest Hearing Aid." CNN, November 9, 2012. *http://www.cnn.com/2012/11/09/ tech/hearing-aid-widex-3d-printing/.*

GE Reports, "Design Freeze Brings Next-Gen LEAP Engine," May 6, 2013.

Gordon, Leslie. "Building Strong but Lightweight Hip Implants." *Medical Design*, October 1, 2008. *http://medicaldesign.com/prototyping/building-strong-yet-lightweight-hip-implants.*

Grimm, Todd. "Stratasys and Oak Ridge National Laboratory Partner to Advance Additive Manufacturing." *Engineering.com,* July 7, 2012. *http://www.engineering.PrintingArticles/ArticleID/4506/Stratasys-and-Oak-Ridge-National-Laboratory-Partner-toAdvance-Additive-Manufacturing.aspx.*

Hessman, Travis. "3D Printing the Supply Chain." *IndustryWeek*, July 15, 2013. *http://mhlnews. com/technology-amp-automation/3-d-printing-sup ply-chain.*

IBISWorld. *3D Printer Manufacturing Market Research Report*, December 2012.

Jurrens, Kevin. "NIST Measurement Science for Additive Manufacturing." National Institute for Standards in Technology. Presentation for PDES, *Inc.,* workshop, March 14, 2013. *http://www.nist.gov/el/msid/infotest/ upload/Jurrens-NIST-AM Projects March2013.pdf.*

LaMonica, Martin. "10 Breakthrough Technologies 2013." MIT Technology Review, April 23, 2013. *http://www.technologyreview.com/featuredstory/ 513716/additive-manufacturing/.*

Lawrence Livermore National Laboratory. "Additive Manufacturing: 3D Printing Metal Parts and Novel Materials." *https://manufacturing.llnl.gov/ additive-manufacturing* (accessed December 3, 2013).

Levy, Stephen. "A Fully Functional 3-D Printed Heart Sooner Than You Think." Qmed, April 15, 2014. *http://www.qmed.com/mpmn/ medtechpulse/fully-functional-3-dprinted-heartsooner-you-think.*

Lind, Michael, and Joshua Freedman. "Value Added: America's Manufacturing Future." New America Foundation, April2012. *http://*

growth.newamerica.net/sites/newamerica.net/files/policydocs/Lind%20 Michael%20and%20Freedman,%20Joshua%20-%20NAF%20-%20Value%20Added%20America%27s%20anufacturing%20Future.pdf.

Lipson, Hod. "Frontiers in Additive Manufacturing." *The Bridge* 42, no. 1 (Spring 2012): 5– 12. *http://www.nae.edu/File.aspx?id=57879.*

Lipson, Hod, and Melba Kurman. *Fabricated: The New World of 3D Printing*. 2013. Hoboken, NJ: Wiley.

Lyons, Brett. "Additive Manufacturing in Aerospace." Presentation for National Academy of Engineering conference, "Frontiers of Engineering 2011: Additive Manufacturing," September 19, 2011. The Manufacturer, "Additive Manufacturing," January 2012, 7.

MarketWired. "Align Technology Announces Second Quarter Fiscal 2013 Results." *http://www.marketwired.com/press-release/align-technology-announces-second-quarterfiscal-2013- results-nasdaq-algn-1812708.htm* (accessed July 7, 2013).

Maxey, Kyle. "3D Printing for the Hearing Impaired." Engineering.com, January 4, 2013. *http:// www.engineering.com/3DPrinting/3DPrinting Articles/ArticleID/ 5128/3D-Printing.*

McKinsey Global Institute. *Disruptive Technologies: Advances That Will Transform Life, Business, and the Global Economy*, May 2013. *http://www.mckinsey.com/insights/business_technology/disruptive_technol ogies.*

———. *Manufacturing the Future: The Next Era ofGlobal Growth and Innovation*, November 2012. *http://www.mckinsey.com/insights/ manufac turing/the future of manufacturing.*

Newcombe, Richard A., Shahram Izadi, Otmar Hilliges, David Molyneaux, David Kim, Andrew J. Davison, Pushmeet Kohli, Jamie Shotton, Steve Hodges, and Andrew Fitzgibbon. "KinectFusion: Real-Time Dense Surface Mapping and Tracking." Microsoft Research, March 2011. *http://research.microsoft.com/pubs/155378/ismar2011.pdf.*

Organization for Economic Cooperation and Development (OECD). STAN Database for Structural Analysis. *http://stats.oecd.org/Index.aspx?DataSet Code=STAN08BIS* (accessed September 24, 2013).

———. OECD *Science, Technology and Industry Outlook 2012. http://dx.doi.org/10.1787/sti_ outlook-2012-en* (accessed June 13, 2013).

Owano, Nancy. "Kor Ecologic Urbee 2 Car Will Move from 3-D Printer to Road." Phys.org, February 2013.

Park, Rachel. "Unsurprisingly, Wohlers Report 2013 Reveals Continued Growth in 3D Printing." 3D Printing Industry, May24, 2013. *http://*

3dprintingindustry.com/2013/05/24/ unsurprisingly-wohlers-report-2013-reveals-continued-growth-in-3d-printing.

Photonics.com. "Pratt & Whitney Opens Additive Manufacturing Center," April 11, 2013. *http://www.photonics.com/Article.aspx?AID=53554.*

Prototype Today. "3D Printer Creates Ear Using Injectable Gels Made of Living Cells." February 28, 2013. *http://www.prototypetoday.com/weill-cornell-medical-college/3dprintercreates-ear-using-injectable-gels-made-of-living-cells.*

Queheillalt, Douglas, and Haydn Wadley. "Pyramidal Lattice Truss Structures with Hollow Trusses." *Materials Science & Engineering* A 397 (February 2005): 132–37. *http://www.virginia.edu/ms/research/wadley/Documents/Publications/Pyramidal_Lattice_Truss_Structure_with_Hollow_Trusses.pdf.*

Quick, Darren. "3D Bio-printer to Create Arteries and Organs." *GizMag*, December 15, 2009. *http://www.gizmag.com/3d-bio-printer/13609.*

Rucks, Gregg. "What Automakers Can Learn from Boeing's Culture of Weight Reduction." *RMI Outlet* (blog), Rocky Mountain Institute, March 28, 2012. *http://blog.rmi.org/blog_what_automakers_canjearn_from_boeings_culture_of_weightjeduction.*

Scott, Justin, Nayanee Gupta, Christopher Weber, Sherrica Newsome, Terry Wohlers, and Tim Caffrey. "Additive Manufacturing: Status and Opportunities." Institute of Defense Analysis. Science and Technology Policy Institute, March 2012. *https://cgsr.llnl.gov/content/assets/docs/IDA_AdditiveM3D_33012_Final.pdf.*

Sealy, Winston. "Additive Manufacturing as a Disruptive Technology: How to Avoid the Pitfall." *American Journal ofEngineering and Technology Research* 12, no. 1 (2012): 86–93. *http://www.ajetr.org/vol12/no1/vol12no110.pdf.*

Selko, Adrienne. "The Next Wave of Manufacturing: Human Organs." *IndustryWeek*, June 6, 2013. *http://www.industryweek.com/technology/next-wave-manufacturing-human-organs.*

Sharma, Ananya. "Interaction with Jim Kor for Sustainability's Brainchild: Urbee 2." *MixFruitSalad*, January 24, 2014. *http://www.mixfruitsalad.com/society/jim-kor-urbee.*

Sirkin, Harold L. "Made in the USA Still Means Something." *Bloomberg Businessweek*, April 10, 2009. *http://www.businessweek.com/managing/content/apr2009/ca20090410 054122.htm.*

Sorrell, Charlie. "Bespoke Innovations Makes Beautiful, Custom Prosthetic Legs." *Wired*, December 22, 2010. *http://www.wired.com/2010/12/bespoke-designs-makes-beautifulcustomprosthetic-legs/.*

Thomas, Douglas. "Economics of the U.S. Additive Manufacturing Industry." U.S. Department of Commerce National Institute of Standards and Technology, June 2013 (prepublication draft).

U.S. Department of Commerce (USDOC). National Institute of Standards and Technology (NIST). "NIST Awards $7.4 Million in Grants for Additive Manufacturing Research." Press release, September 19, 2013. *http://www.nist.gov/director/3dgrant.cfm.*

U.S. Department of Energy. Advanced Manufacturing Office. "Additive Manufacturing: Pursuing the Promises," August 2012. *http://www1.eere.energy.gov/manufacturing/pdfs/ additive_manufacturing.pdf.*

West, Karl. "Melted Metal Cuts Plane's Fuel Bill." *Sunday Times* (London), February 13, 2011. *http://www.thesundaytimes.co.uk/sto/business/energy and environment/article547163.ece.*

White House. "State of the Union Address," January 24, 2012. *http://www.whitehouse.gov/thepress-office/2012/01/24/remarks-president-state-union-address.*

Wohlers, Terry. "*Wohlers Report 2012*: Additive Manufacturing and 3D Printing State of the Industry." Wohlers Associates, Inc., 2012.

Wohlers, Terry, and Tim Caffrey. "Additive Manufacturing: Going Mainstream." *ManufacturingEngineeringMedia.com,* June 2013, 67–73. *http://www.sme.org/uploadedFiles/Publications/ME_Magazine/2013/June_2013/June%202013%20f1%20Additive.pdf*(accessed July 2, 2013).

World Economic Forum (WEF). *The Global Competitiveness Report, 2010–2011*. WEF, 2010. *http://www3.weforum.org/docs/WEF GlobalCompetitivenessReport 2010-11.pdf.*

End Notes

[1] This article represents solely the views of the author and not the views of the United States International Trade Commission or any of its individual Commissioners. This paper should be cited as the work of the author only, and not as an official Commission document. The author thanks Dylan Carlson and the JICE board for their contributions.

[2] USDOE, "Advanced Manufacturing: Pursuing the Promises," August 2012, 1.

[3] Wohlers, "Additive Manufacturing," June 2013, 67–73.

[4] Park, "Unsurprisingly, Wohlers," May 24, 2013, 1.

[5] For the purposes of this article, "automotive" is used interchangeably with "motor vehicle," and "medical" encompasses "dental."

[6] Sealy, "Additive Manufacturing as a Disruptive Technology," 2012, 92.

[7] Rapid prototyping and additive manufacturing differ according to product characteristics. While prototypes are used to show special product properties or functions during the product development phase, additive manufacturing also delivers final products.

[8] Approximately 28 percent of additively manufactured parts are final products. Wohlers, "Additive Manufacturing," June 2013, 7–12. Desktop computers and industrial lasers have facilitated advancements in additive manufacturing.

[9] OECD, Structural Analysis Database (accessed September 24, 2013).

[10] Queheillalt and Wadley, "Pyramidal Lattice Truss Structures," 2005, 132–37; A truss that resembles latticework because of diagonal placement of members connecting the upper and lower chords.

[11] Subtractive manufacturing processes such as machining take raw material and remove and shape it into the desired final form. In some cases, over 90 percent of a billet of raw material may be removed before the product is finished. Campbell and Slotwinski, "Metrology for Additive Manufacturing," 2013, 154.

[12] Matthews International Corporation representative, interview by author, March 5, 2014.

[13] White House, "State of the Union Address," January 24, 2012.

[14] A polymer is a chemical compound made of small molecules that are arranged in a simple repeating structure to form a larger molecule. Merriam-Webster Dictionary website, http://www.merriamwebster.com/dictionary/polymer (accessed May 21, 2014). Examples of polymers often used in additive manufacturing are polycarbonate and high-density polyethylene.

[15] Material extrusion is the selective dispensation of material through a nozzle or orifice. Wohlers, Wohlers Report 2012, 2012, 68.

[16] Material jetting is the selective deposition of droplets of build material via ink-jet print head nozzles. Wohlers, Wohlers Report 2012, 2012, 68.

[17] Binder jetting is the selective deposition of a liquid bonding agent through ink-jet print head nozzles to join powder materials. Wohlers, Wohlers Report 2012, 2012, 68.

[18] Sheet lamination involves bonding sheets of material together. Wohlers, Wohlers Report 2012, 68.

[19] Vat photopolymerization is selective curing of liquid photopolymer (light-sensitive polymer) in a vat via light-activated polymerization. Wohlers, Wohlers Report 2012, 2012, 68.

[20] Powder bed fusion is selective fusing of powder bed regions via thermal energy. Wohlers, Wohlers Report 2012, 2012, 68.

[21] Directed energy deposition is simultaneous fusion and deposition of material. Wohlers, Wohlers Report 2012, 2012, 68.

[22] A list of process category names and definitions which the ASTM International Committee F42 on Additive Manufacturing Technologies approved in January 2012 appears in the appendix; Additive manufacturing machines range in size from desktop models, which weigh just a few pounds, to an Objet-Stratysys model, which measures 9 feet x 6.7 feet x 5.5 feet and weighs 6,325 pounds.

[23] Wohlers, Wohlers Report 2012, 2012, 83.

[24] Wohlers, Wohlers Report 2012, 2012, 74, 82.

[25] Georgia Institute of Technology Representative, interview by author, June 25, 2013.

[26] Wohlers, Wohlers Report 2012, 2012, 183. When machine tooling is not needed and the restrictions of design for manufacture and assembly are eliminated, the possibilities for design are limited only by the available tools and a designer's imagination.

[27] Vacari, "3D Printing Industry," April 19, 2013, 3.

[28] In traditional manufacturing, the prototype, tooling, and setup processes are too time-consuming and expensive to profitably produce small volumes.

[29] Campbell et al., "Could 3D Printing Change the World?," October 2011, 6. If production is decentralized, then hundreds of thousands of units of a product may be produced at multiple locations that are near the source of demand, rather than at one location.

[30] Plastic materials for additive manufacturing can be 10 to 50 times as expensive as plastic materials for injection molding.

[31] OECD, Structural Analysis Database (accessed September 24, 2013).

[32] Wohlers, Wohlers Report 2012, 2012, 133.

[33] Other countries with significant stocks of industrial additive manufacturing machines include Japan (10.2 percent of global total), Germany (9.3 percent), and China (8.6 percent). Wohlers, Wohlers Report 2012, 2012, 18.

[34] This paper focuses on the second-, third-, and fourth-largest sectors because they are relatively distinct. It does not address the single largest sector, consumer products and electronics, because it is too varied and diffuse to discuss comparably. Approximately 20.3 percent of additive manufacturing is within consumer products and electronics. Wohlers, Wohlers Report 2012, 2012, 18.

[35] Wohlers, Wohlers Report 2012, 2012, 18.

[36] Thomas, Economics ofthe U.S. Additive Manufacturing Industry (prepublication draft), June 2013, 38.

[37] Thomas, Economics of the U.S. Additive Manufacturing Industry (prepublication draft) June 2013.

[38] Boulton, "Barbie, Auto Parts Hot Off the Press," June 6, 2013, B5.

[39] Build volume refers to the external dimensions.

[40] Brooks, "Auto Giant Harnesses New Super-scale Laser," November 15, 2012, 7.

[41] Sharma, Ananya, "Interaction with Jim Kor," January 24, 2014.

[42] Owano, "Kor Ecologic," February 28, 2013, 1.

[43] Owano, "Kor Ecologic," February 28, 2013, 1.

[44] Wohlers, Wohlers Report 2012, 2012, 18.

[45] Campbell, Bourell, and Gibson, "Additive Manufacturing," 2012, 258.

[46] Thomas, Economics of the U.S. Additive Manufacturing Industry (prepublication draft), June 2013.

[47] A malocclusion is a misalignment of teeth or incorrect relation between the teeth of the two dental arches.

[48] MarketWired, "Align Technology Announces," April 18, 2013, 1.

[49] Prototype Today, "3D Printer Creates Ear Using Injectable Gels," February 28, 2013.

[50] Keith Murphy Selko, CEO at Organovo, IndustryWeek, "The Next Wave of Manufacturing," June 6, 2013, 3.

[51] Lipson, "Frontiers in Additive Manufacturing," Spring 2012, 1.

[52] BoneZone, "The Future of Additive Manufacturing," March 5, 2013.

[53] Wohlers, Wohlers Report 2012, 2012, 18.

[54] Rucks, "What Automakers Can Learn from Boeing's Culture," March 28, 2012.

[55] Lyons, "Additive Manufacturing in Aerospace," September 19, 2011, 2; Campbell et al., "Could 3DPrinting Change the World?," October 2011, 1; Campbell and Slotwinski,

"Metrology for Additive Manufacturing," 2013, 154. Traditional manufacturing and subassembly of 16 parts, plus glue, of an air duct for an F-18 fighter jet is additively manufactured into a component. It is also optimized for air flow efficiencies.

56 A reduction of one kilogram in the weight of an airliner saves about $3,000 annually in fuel and reduces carbon emissions. Additive manufacturing could help build greener aircraft—especially if all of the 1,000 or so titanium parts in an aircraft could be printed. Economist, "3D Printing," February 14, 2011, 19.

57 Lyons, "Additive Manufacturing in Aerospace," September 19, 2011, 9.

58 Boeing.com, "Boeing Launches New Manufacturing Venture," September 3, 2002.

59 Freedman, "Layer by Layer, December 19, 2011.

60 LaMonica, "10 Breakthrough Technologies 2013, April 23, 2013, 2.

61 Davidson, "3-D Printing Could Remake U.S. Manufacturing, June 10, 2013, 3.

62 Davidson, "3-D Printing Could Remake U.S. Manufacturing, June 10, 2013, 3.

63 Industry representative, interview by author, July 15, 2013.

64 Industry representative, interview by author, July 15, 2013.

65 Engineering.com, "3D Printed Rocket Blasts Off, July 15, 2013.

66 Industry representative, interview by author, July 11, 2013.

67 Castle Island, "Additive Manufacturing: What RP Will Be, n.d. (accessed May 1, 2014).

68 Hessman, "3D Printing the Supply Chain," July 15, 2013, 2.

69 Wohlers, Wohlers Report 2012, 2012, 13.

70 Brooke, "Chevrolet Uses Rapid Prototyping," June 11, 2013.

71 Brooke, "Chevrolet Uses Rapid Prototyping," June 11, 2013.

72 Distributed manufacturing, also known as distributed production and local manufacturing, is a coordinated networkof geographically dispersed manufacturing facilities.

73 Sorrell, "Bespoke Innovations Makes Beautiful, Custom Prosthetic Legs," December 22, 2010, 1.

74 USDOE, "Additive Manufacturing: Pursuing the Promises," August 2012, 2.

75 IBISWorld, 3D Printer Manufacturing Market Research Report, December 2012, 23.

76 The Manufacturer, "Additive Manufacturing," January 2012, 7.

77 Industry representative, interview by author, July 11, 2013. Body side cladding is a protective composite paneling that covers the lower part of the doors and fenders around the wheels.

78 BMW North America website, "Build your Own," http://www.bmwusa.com/ (accessed July 8, 2013).

79 Scott et al., "Additive Manufacturing," March 2012, 10.

80 Glass, "Pitch Perfect: The Quest to Create," November 9, 2012; Maxey, "3D Printing for the Hearing Impaired," January 4, 2013.

81 Wohlers, Wohlers Report 2012, 2012, 187.

82 Campbell and Ivanova, "3D Printing for the Hearing Impaired" August 2013, 8, 119-120.

83 In conformal electronics, protective material is applied to electronic circuitry.

84 De Beer, "Additive Manufacturing: Turning Mind into Matter," May 31, 2013, 17.

85 Scott et al., "Additive Manufacturing," March 2012, 8.

86 Federation of European Screen Printers Associations, "3D Printing, Additive Manufacturing and Drivers for Adoption," October 2012, 9.

87 Lawrence Livermore National Laboratory, "Additive Manufacturing: 3D Printing Metal Parts and Novel Materials," n.d. (accessed December 3, 2013).

88 Scott et al., "Additive Manufacturing," March 2012, 10.

89 Newcombe, "KinectFusion: Real-Time Dense Surface Mapping and Tracking," March 2011, 9.

90 Scott et al., "Additive Manufacturing," March 2012, 11.

[91] BioPortfolio, "Additive Manufacturing Techniques for Producing Tissue," July 22, 2013.

[92] Levy, "A Fully Functional 3-D Printed Heart," April 15, 2014.

[93] Slotwinski, "Additive Manufacturing at NIST," March 9, 2013, 2.

[94] Bello, "NIST Awards $7.4 Million in Grants for Additive Manufacturing Research," September 2013 (accessed October 8, 2013).

[95] Dehoff, "Thinking Beyond Today," July 11, 2013.

[96] In January 2013, the Brookings-Rockefeller Project on State and Metropolitan Innovation put NAMII on its annual "Top 10 State and Metropolitan Innovations to Watch" list, recognizing its potential to create jobs, grow regional economies, and boost global competitiveness.

[97] America Makes website, https://americamakes.us/ (accessed June 13, 2013).

[98] America Makes website, https://americamakes.us/ (accessed June 20, 2013).

[99] America Makes website, https://americamakes.us/ (accessed June 13, 2013).

[100] America Makes website, https://americamakes.us/ (accessed June 20, 2013).

[101] The United States dominated basic research for additive manufacturing through the late 1990s, when public funding decreased. Today, Europe, especially Germany, dominates applied research in the technology. Industry representative, interview by author, July 15, 2013.

[102] EWI is a member organization which develops and applies manufacturing technology innovation within the manufacturing industry. EWI, Additive Manufacturing Consortium website, http://ewi.org/additivemanufacturing-consortium (accessed June 20, 2013).

[103] Fused deposition modeling heats thermoplastic material to a semi-liquid state and extrudes it according to computer-controlled paths. The resulting parts are unrivaled in mechanical, thermal, and chemical strength.

[104] Grimm, "Stratasys and Oak Ridge," July 7, 2012.

[105] Photonics.com, "Pratt & Whitney Opens Additive Manufacturing Center," April 11, 2013; Aerojet Rocketdyne acquired Pratt & Whitney Rocketdyne in June 2013.

[106] Sealy, "Additive Manufacturing as a Disruptive Technology," 2012, 92.

[107] Scott et al., "Additive Manufacturing: Status and Opportunities," March 2012, 20.

[108] Scott et al., "Additive Manufacturing: Status and Opportunities," March 2012, 20.

[109] The adoption of parts made via additive manufacturing processes into critical applications, such as aerospace engine components, is hampered by a lack of consensus (e.g., regarding material properties for additive manufacturing) required for many procurement specifications; the NIST Materials Data Program provides evaluated data on phase equilibria, structure and characterization, and performance properties.

[110] Industry representative, interview by author, July 15, 2013.

[111] Property data summaries are collections of property values derived from surveys of published data. These collections typically focus on either one material or one particular property. Studies of specific materials typically include thermal, mechanical, structural, and chemical properties, while studies of particular properties survey one property across many materials. The property values may be typical, evaluated, or validated. Values described as typical are derived from values for nominally similar materials. Jurrens, "NIST Measurement Science for Additive Manufacturing," March 14, 2013, 7.

[112] Campbell and Slotwinski, "Metrology for Additive Manufacturing," 2013, 164.

[113] Campbell and Slotwinski, "Metrology for Additive Manufacturing," 2013, 163.

[114] Additive Manufacturing Institute, "Manufacturing Engineering," January 5, 2013, 16.

[115] NIST representative, interview by author, June 25, 2013.

[116] Georgia Technical Institute representative, interview by author, June 25, 2013.

[117] Characterization in materials science refers to the use of external techniques to probe the internal structure and properties of a material (e.g., testing or analyzing a material to visualize its internal structure and gain knowledge about the distribution and interaction of its elements).

[118] The most common additive manufacturing materials are metals and plastics and/or polymers. Various filled and composite materials are also used, as well as ceramic and ceramic-metal hybrids. The most common traditional manufacturing materials are metal, ceramics, and polymers. The Library of Manufacturing website, http://thelibraryofmanufacturing.com (accessed May 21, 2014).

[119] The versatility of what can be made will grow exponentially with the increasing number of primary materials that can be simultaneously mixed and printed. Hybrids are not just mixed materials; they are new kinds of materials with unique properties.

[120] Bourell, Leu, and Rosen, Roadmapfor Additive Manufacturing, 2009, 31.

[121] Wohlers, Wohlers Report 2012, 2012, 78; Campbell and Slotwinski, "Metrology for Additive Manufacturing," 2013, 167. The layer-wise nature of various additive manufacturing processes can lead to unwanted levels of porosity, especially for processes that involve melting of input materials. Porosity is beneficial for some biomedical implants, but can cause metal aerospace parts to fail.

[122] McKinsey Global Institute, Manufacturing the Future, 2013, 90.

[123] Gordon, "Building Strong but Lightweight," October 1, 2011, 7.

[124] The average industrial additive manufacturing machine now sells for about $75,000, and some machines cost more than $1 million. McKinsey & Co., Disruptive Technologies, 2013, 109.

[125] Limitations vary by additive manufacturing technique.

[126] Minimum wall thickness is the absolute minimal thickness of a pipe or structure to contain its contents.

[127] Deloitte, "TNT Predictions," March 17, 2012, 16.

[128] Bourell, Leu, and Rosen, Roadmap for Additive Manufacturing, 2009. This document was designed with the goal of developing and articulating a systematic plan for additive manufacturing research for the next 10–12 years. Other sources outlined the same barriers. In 2013, America Makes announced plans to develop a national additive manufacturing roadmap.

[129] Rapid prototyping is quicker with additive manufacturing.

[130] Bourell, Leu, and Rosen, Roadmapfor Additive Manufacturing, 2009, 16.

[131] Neither category includes secondary processes, such as making castings from this tooling. This market segment grew 25.1 percent, to $1 billion in 2011, up from $836 million in 2010, when it grew 27.3 percent. The combined total for overall additive manufacturing products and services—both primary and secondary markets— was $2.8 billion, an increase of 27.4 percent from $2.2 billion generated in 2010. Wohlers, Wohlers Report 2012, 2012, 130.

[132] Wohlers, Wohlers Report 2012, 2012, 125.

[133] Wohlers Associates considers and reports the secondary market segment separately.

[134] Wohlers, Wohlers Report 2012, 2012, 126. The average selling price of an industrial additive manufacturing machine was $73,220 in 2011; Wohlers, Wohlers Report 2012, 2012, 127.

[135] Wohlers, Wohlers Report 2012, 2012, 130.

[136] Wohlers, Wohlers Report 2012, 2012, 131.

[137] IBISWorld, "3D Printer Manufacturing Market Research Report," December 2012, 18.

[138] Wohlers, Wohlers Report 2012, 2012, 131.

[139] Wohlers, Wohlers Report 2012, 2012, 137.

[140] Metals, plastics, and other feedstock require special processing (e.g., polymerization) for use in additive manufacturing.

[141] Wohlers, Wohlers Report 2012, 2012, 79.

[142] McKinsey Global Institute, Disruptive Technologies, 2013, 109.

[143] Rilsan Invent Natural and Rilsan Invent Black. Wohlers, Wohlers Report 2012, 2012, 76.

[144] Ceramic matrix composites handle enormous stress and temperatures as high as 2,400 degrees Fahrenheit inside gas turbines and jet engines, and make them much more fuel efficient. CFM estimates that the reduced weight of the additively manufactured components will reduce fuel consumption by about 15 percent—enough to save nearly $1 million per year per airplane, assuming a fuel cost of $2.50 per gallon. Destefani, "Ceramic Matrix Composites Make Inroads in Aerospace," May 14, 2013, 1.

[145] GE Reports, "Design Freeze Brings Next-Gen LEAP Engine," May 6, 2013, 2.

[146] In situ sensors are in constant contact with the medium they are measuring. For example, a spectroscopic sensor detects and categorizes defects, predicts the composition and phase transformation of the medium, and monitors manufacturing quality in real time.

[147] Objet Connex website, http://www.stratasys.com/3d-printers/design-series/precision/objet-connex500 (accessed July 2, 2013).

[148] Wohlers, Wohlers Report 2012, 2012, 22.

[149] Wohlers, Wohlers Report 2012, 2012, 199.

[150] Lipson and Kurman, Fabricated, 2013, 221.

[151] Bourell, Leu, and Rosen, Roadmapfor Additive Manufacturing, 2009, 16.

[152] RepRap website, http://reprappro.com/about (accessed July 2, 2013).

[153] Selective laser sintering involves the use of a high-powered laser (for example, a carbon dioxide laser) to fuse small particles of plastic, metal (direct metal laser sintering), ceramic, or glass powders into a mass that has a desired three-dimensional shape.

[154] Wohlers, Wohlers Report 2012, 2012, 253. Selective laser sintering can easily make very complex geometries directly from digital CAD data. While it began as a way to build prototype parts early in the design cycle, it is increasingly being used in limited-run manufacturing to produce end-use parts.

[155] Thomas, Economics of the U.S. Additive Manufacturing Industry (prepublication draft), June 2013, 2.

[156] CSC, "3D Printing and the Future of Manufacturing," Fall 2012, 21.

[157] CSC, "3D Printing and the Future of Manufacturing," Fall 2012, 21.

[158] Economist, "Print Me a Stradivarius," April 21, 2012, 17.

[159] Baily, Manyika, and Gupta, "U.S. Productivity Growth: An Optimistic Perspective," Spring 2013, 8.

[160] As noted, among the countries for which OECD STAN data are available, for example, the United States also has the largest R&D expenditure for total manufacturing growth.

INDEX

D

E

R

S

T

U

V

W

Y

Z

com/3DPrinting/3D